SCIENCE AND IMAGINATION

By Marjorie Nicolson

ARCHON BOOKS

1976

First published 1956

by Cornell University Press as a Great Seal Book

Reprinted 1976 with permission of Cornell University Press

in an unabridged and unaltered edition as an Archon Book,

an imprint of The Shoe String Press, Inc.,

Hamden, Connecticut 06514

Library of Congress Cataloging in Publication Data

Nicolson, Marjorie Hope, 1894–
    Science and imagination.

    Reprint of the 1962 ed. published by Great Seal Books, Ithaca, New York.
    Includes bibliographical references and index.
    CONTENTS: The telescope and imagination.—The "new astronomy" and English imagination.—Kepler, the Somnium, and John Donne. [etc.]
    1. Literature and science—Addresses, essays, lectures. 2. English literature—History and criticism—Addresses, essays, lectures. I. Title.
PR 149.S4N5   1976   820′.9′31   76–11730
ISBN 0–208–01603–1

Printed in the United States of America

Preface

WHEN these essays began to appear in 1935, the approach to literature through the history of science seemed much more novel than it does today. Carson Duncan's *New Science and English Literature*, a pioneer work in the field, had appeared as early as 1913; unfortunately it was privately printed and never gained the audience it deserved. I myself did not see it until several of these articles had been printed. Claude Lloyd's "Shadwell and the Virtuosi" appeared in the *Publications of the Modern Language Association* in 1930. Richard F. Jones had published several of his articles on the influence of science on prose style and, still earlier, his important *Background of the Battle of the Books,* in which he proved that the English "Battle" was no pale reflection of a French literary quarrel but had its roots in a native English scientific tradition. With the exception of Jones's studies, however, publications in the field of science and literature had been sporadic. Shortly after the earliest of my papers appeared, Charles Monroe Coffin published *John Donne and the New Philosophy,* in which he treated definitively the subject I had handled much more superficially in the second of the present essays. In 1937 appeared Francis Johnson's *Astronomical Thought in Renaissance England,* a work of major importance to all later workers in the field. Thanks in large part to Professors Coffin and Johnson, the relation of science to literature became a major field for research. During the last two decades interest has developed so markedly that a "Literature and Science" group has been established by the Modern Language Association, and the bibliography sponsored by it has grown to major proportions.

The methodology I followed in these studies is that which has come to be called "the history of ideas," less familiar twenty years ago than it is today. Since literary historians have frequently asked me about my "influences" in the development of this method, I shall try to indicate my chief sources, since it is always difficult to determine the origin of any new movement in scholarship. To a greater extent than we may have realized, many of us, I feel sure, were influenced by Alfred North Whitehead's *Science and the Modern World*, particularly by the brilliant chapter on "The Century of Genius." Professor Whitehead was not particularly concerned with literature, but that even his passing remarks fell upon fertile soil may be seen in such a study as Carl Grabo's *Newton among Poets*, in which Mr. Grabo took his point of departure from a sentence in which Whitehead suggested that if Shelley had not chosen the life of a poet, he might have been a great chemist. While I cannot recall any particular sentence of Whitehead's that led me to my present interest, I am sure of his influence. Even more important in leading me to the subject matter and the methodology of these articles was my academic acquaintance with Professor Arthur Lovejoy. His monumental *Great Chain of Being* appeared in 1935, when my earlier essays had been completed, but I had had the privilege of working with him at the Johns Hopkins University from 1923 to 1926. While I was too early to hear his lectures on the chain of being, I was among the group of students who worked with him on "the discrimination of the Romanticisms" and on the concept of Nature. His methodology, his study of the impact of philosophy upon literature—or rather of the reflection in both philosophy and literature of the same currents of thought—influenced every student who studied with him, none more than myself. Preparation for the editing of the *Conway Letters,* which I published in 1930, had already led me to read in the history of medicine. However, it was largely the stimulus I had received from Professor Lovejoy that caused me to read more deeply in the history of science, with the idea of attempting to contribute to the study of the interplay of science and literature something of the sort he had contributed so richly to the study of the interrelationship of literature and philosophy.

For inclusion in the present volume, I have selected chiefly articles dealing with the telescope and the microscope, since the effect of these new instruments upon literary imagination was almost immediate. For centuries astronomy had been the most familiar science to laymen: knowledge of the constellations was important to the traveler, whether by land or sea, and knowledge of astrology was essential to both physician and patient in the diagnosis and treatment of disease. While the Copernican hypothesis was known to the intelligent reader shortly after its publication, it did not fundamentally change his attitude toward the long-accepted Ptolemaic theory, since one explained the "appearances" as well as the other, since the new theory did not challenge the astronomical teachings about the planets, and since—no matter what the mathematician or the philosopher might say about the relative position of the earth and sun—the "little world of man," firmly anchored beneath his feet, remained subjectively the center of the universe, as indeed it does to most of us today. But the telescopic discoveries of Galileo, "the man who saw through heaven," mattered profoundly, as I have tried to show. The upheaval in human thought produced by the "new astronomy" is the subject of the first essays in this volume. The microscope, a natural development from the telescope, was even more readily intelligible to the layman. Its effect upon imagination is found in both prose and poetry.

I have not included in this volume various other essays in which I have studied the effect of the new science upon literature—for example, the monograph *A World in the Moon,* an article on "Swift's Flying Island," and another on "The First Electrical Flying Machine"—since much of the material was included in my *Voyages to the Moon* (New York, 1948). Some ideas suggested in passing in the present volume I have since developed further in *Newton Demands the Muse* (Princeton, 1946) and *The Breaking of the Circle* (Evanston, 1950).

The articles are republished largely as they were written, with occasional verbal changes, the addition of a few paragraphs to "The New Astronomy and English Literary Imagination," and some compression of "Kepler, the *Somnium,* and John Donne." The original footnotes have been greatly reduced in number and length, since many of the references may conveniently be found in the volumes of Professors Coffin and Johnson. I shall not repeat the acknowledgments of assistance I made earlier, but I must again express my gratitude to Professor Nora Mohler of the Department of Physics, Smith College, who collaborated with me on the two articles on *Gulliver's Travels,* and whose name appears here as co-author of "The Scientific Background of Swift's *Voyage to Laputa.*" In addition, I wish to express my appreciation to the editors and publishers of the various periodicals in which these essays originally appeared: the University of Chicago Press, the Johns Hopkins Press, the University of North Carolina Press, the College of the City of New York, the Trustees of Smith College, and the Royal Society of London.

M. H. N.

Columbia University
August 1956

Contents

I. The Telescope and Imagination

DURING the last few years—perhaps because of the dominant interest of our own time—students of literary history have become aware of the importance of the scientific background in determining the direction of certain currents of literature, and have been increasingly conscious of the extent to which major and minor writers have felt the pressure of contemporary scientific conceptions. Of all the periods in which scientific thought has transformed the world, no age until our own saw more tremendous changes than that of the Renaissance. The most epoch-making of these changes came about through what is vaguely called "Copernicanism." The difference between "old" and "new" is shown in the difference between generations who felt their earth the center of the universe and generations who learned that their earth is no such thing. Yet, though we have paid lip-service to that theory, we have felt our convictions fall before poet after poet who, knowing with his intellect the hypotheses of Copernicus, still felt that the "little world of man" remains a solid ball beneath his feet, the center of his universe. The student of seventeenth-century literature who reads thoughtfully those earlier poets who first experienced the strangeness of the "new astronomy," and the later poets who accepted it as a matter of course, be-

comes aware that there was little stirring of the cosmic imagination even among those who defended Copernicus. Ultimately
he reaches the conclusion that, although the intelligent layman of the seventeenth century was aware of the so-called
Copernican hypothesis, in itself the hypothesis disturbed him
little; it led in few cases to either optimism or pessimism, but
rather, as in the case of Milton's angel, to a judicial weighing
of hypotheses without too much concern as to which should
finally be proved true.[1]

Yet something in the "new astronomy" led to both optimism
and pessimism. There is a feeling here of change, of awareness
of astronomical implication which both disturbs and fascinates
the seventeenth-century mind. On the one hand, man is shrinking back from an unknown gulf of immensity, in which he
feels himself swallowed up; on the other, he is, like Bruno,
"rising on wings sublime" to a spaciousness of thought he had
not known before. The poetic and religious imagination of the
century was not only influenced, but actually changed, by
something latent in the "new astronomy." New figures of
speech appear, new themes for literature are found, new attitudes toward life are experienced, even a new conception of
Deity emerges. These have little to do with the problem of the
relative position of the earth and sun; they are not even, for
the most part, the consequence of man's knowledge that his
earth is not a special creation of God's, the center of the universe. The century was aware less of the position of the world
than of the immensity of the universe, and the possibility of a

[1] I suspect that the opinion of the second Viscount Conway on the subject was characteristic of many an intelligent gentleman who, like himself,
was "noe otherway a Scholer then a Scotch pedler is a Marchant." He
wrote to his daughter-in-law in 1651 (*Conway Letters* [New Haven, 1930],
p. 32): "Copernicus hath divers followers not bycause his opinion is true
but bycause the opinion is different from what all men in all ages ever
had, for he hath not proved that there is any ill consequence by holding
that the Earth doth stand still and the heavens move, or discover the least
Error in this Tenent, but only he hath very ingeniously shewed that it
may be as well demonstrated that the heavens stand still as that the earth
stands still, we shall know no more then we doe if we think as he doth."

2

plurality of worlds. It is this which troubles and enthrals; the solid earth shrinks to minute proportions as man surveys the new cosmos; it is a tiny ball, moving in indefinite space, and beyond it are other worlds with other suns, all part of a cosmic scheme defeating imagination. Not only that; the seventeenth century, as it became conscious of indefinite space, became aware also that in the little world a new microcosm reflected the new macrocosm. Again and again we find men turning, as Pascal in the most magnificent passage on the subject, from contemplating "entire nature in her height and full majesty," from a bewildering universe in which "this visible world is but an imperceptible point in the ample bosom of nature," to "another prodigy equally astonishing," the world "of things most minute" in which the "conceivable immensity of nature" is displayed again "in the compass of this abbreviation of an atom." Like Pascal, many a man after the invention of the microscope viewed "therein an infinity of worlds" and lost himself "in these wonders, as astonishing in their littleness as the others in their magnitude." [2]

Such passages as these are not the direct result of Copernicanism, but they are none the less the result of the new astronomy. Not Copernicus but Bruno, Galileo, Kepler are reflected here. On the one hand, the poetic intuition of the Nolan; on the other, actual sense experience led Milton, Pascal, and others of the midcentury to an awareness of the vastness and the minuteness of the universe and of man—an experience which in our age is so much taken for granted that only with difficulty can we think ourselves back to the situation of that first generation of men who by means of the telescope, and of its descendant, the microscope, sometimes in one night saw the crashing-down of the "flaming ramparts of the world" when

> They viewed the vast immeasurable Abyss,
> Outrageous as a sea, dark, wasteful, wild.

In this article, and in those which follow, I shall seek to trace that experience, attempting to see how the imagination of first

2 Blaise Pascal, *Thoughts*, trans. O. W. Wight (1893), pp. 158–60.

one, then another, was transformed as he looked through the telescope, then through the microscope, and as he read the volumes of others who had done so. We may perhaps date the beginning of modern thought from the night of January 7, 1610, when Galileo, by means of the instrument which he had developed, thought he perceived new planets and new worlds. But there were two earlier moments which must be considered in any study of the awakening of modern astronomical imagination—differing in degree, but much the same in kind—discovery of Tycho's "new star" of 1572 and of Kepler's of 1604. These were the precursors that prepared men's minds to accept what would otherwise have seemed incredible in Galileo's *Sidereus Nuncius* in 1610—the most important single publication, it seems to me, of the seventeenth century, so far as its effect upon imagination is concerned. I shall try at the present time to reconstruct the instantaneous effect the discoveries reported there had upon the poets of Galileo's own country, in order that their effect upon poetic imagination in England may be better understood. In the papers that follow, I shall trace the course of that effect in England, the merging of Galilean ideas with those already native there, and then shall follow the development of the telescope and microscope as they appear in literature, watching new figures of speech, new literary themes, new cosmic epics, most of all the transformation of poetic and religious imagination by ideas which, once grasped, man has never been able to forget.

I

The first of the great discoveries which were to make the thoughtful layman of the day aware of the new heavens and the new earth occurred on November 11, 1572, when Tycho Brahe, the Danish astronomer at Uraniborg, noticed in the constellation of Cassiopeia what seemed an unfamiliar bright star. Astonishment seized him, and at first he not unnaturally believed that he must be mistaken; for, according to accepted Aristotelianism, the stars, even more certainly than the sparrows, were known and numbered. In the classic doctrine of the schools, the heavens were perfect and immutable, subject neither to change nor decay, the heavenly galaxy unalterable. As Tycho continued to observe the phenomenon, however, the brilliance of the star—it was equal to Venus at its brightest—and its continuance—it did not fall below the first magnitude for four months—assured him that he could not have been mistaken in his observations. He was ultimately prevailed upon to write and publish a treatise in which he maintained stoutly that this could not be, as many thought, a comet, and established to the satisfaction of those not too closely bound by orthodoxy that this was a *nova*. Such a contention was not only, of course, in opposition to the orthodox philosophy of the heavens, but—much more important to the average mind —it struck at the root of established astrology. Though others saw and studied the new light, "Tycho's star" it was, and "Tycho's star" it remained in popular imagination during the century. As late as 1650, indeed, long after Kepler's new star of 1604, long after *novae* had come to be generally accepted, Dryden went back to Tycho for a figure of speech, when he wrote on the death of Lord Hastings:

> Liv'd Tycho now, struck with this ray (which shone
> More bright i' th' morn, than others' beam at noon)
> He'd take his *astrolabe*, and seek out there
> What new star 'twas did gild our hemisphere.[3]

Had Tycho lived three years longer, he might have had opportunity to see the reception in 1604 of the news of another new star, and to realize that his seed had not fallen upon entirely barren soil. But Tycho died in 1601; and Kepler, with whom the discovery of the new star of 1604 is associated, had been only an infant in 1572. Yet Kepler had had opportunity to hear at first hand Tycho's theories on new stars, and the appearance of the star of 1604 could hardly have been as startling to him as that of 1572 had seemed to Tycho. In 1599 when Tycho, who had left Uraniborg after many years of astronomical observation, had settled at Prague under the patronage of

3 *Upon the Death of Lord Hastings*, ll. 43–46.

the Emperor Rudolph, he learned of a promising and daring young mathematician, Johann Kepler, who was greatly in need of support. Kepler was at this time but twenty-eight years of age, though his intellectual experiences had been those of a man twice his years. As early as his university days Kepler had come under the spell of Copernicanism, even then beginning to be a dangerous heresy.[4]

In 1596 Kepler published the first of his great works, the *Mysterium Cosmographicum*,[5] which, less important in the history of thought than his later works, seems to have remained his own favorite, for Kepler remained, in his own mind, first a mystic, second a scientist. The *Mysterium Cosmographicum* appealed profoundly to that type of Renaissance mind which, like Sir Thomas Browne, loved mystical mathematics. Here one may find the mysteries of numbers and of distances, the conception of the Great Geometer who in his design of the universe has been moved by what Sir Thomas Browne called the "Quincuncial Ordination" in the "strange Cryptography" of "his starrie Booke of Heaven." While there is much, even in the *Mysterium Cosmographicum*, which contributed to the history of astronomy as the scientist knows it, it is essential for any understanding of the effect which Kepler was to produce upon the imaginative minds of later poets and theologians to realize that his first public defense of Copernicus was based upon his profound belief that Copernicanism was ultimately consistent with mysticism.

[4] While the Ptolemaic system was publicly expounded at Tübingen, where at the age of seventeen Kepler received the Bachelor's degree, Michael Maestlin, professor of mathematics who taught the accepted philosophy in his lectures, took occasion to instruct this most promising of his pupils privately in Copernican principles. Cf. C. Carl Rufus, "Kepler as an Astronomer" in *Johann Kepler, 1571–1630: A Tercentenary Commemoration of His Life and Works . . . prepared under the Auspices of the History of Science Society* (1931), pp. 4–5.

[5] *Prodromus Dissertationum Mathematicarum continens Mysterium Cosmographicum, . . . Addita est Narratio G. Joachimi Rhetici De Libris Revolutionum, atque Admirandis de Numero, Ordine, et Distantiis Sphaerarum Mundi Hypothesibus . . . N. Copernici* (Tubingae, 1596).

6

The *Mysterium Cosmographicum* was followed by a series of short papers, all marked by the same mysticism, which, after his banishment to Hungary, Kepler sent back to Maestlin in Tübingen, with a pitiful request for aid. Perhaps as a result, in 1599 Tycho urged Kepler to come to Prague "as a welcome friend," and after some hesitation on Kepler's part, the offer was accepted. For a little more than a year the two men worked together, finding fundamental agreement in spite of their differences, and at Tycho's sudden death the following year Kepler found himself heir not only to Tycho's post of chief imperial mathematician but—what was to him of far greater importance—spiritual heir to Tycho's great legacy of astronomical observations and hypotheses, which were to affect all his later discoveries and conclusions. The first of his important publications after Tycho's death—concerning the new star— showed clearly his familiarity with Tycho's long ponderings on that earlier star of 1572.

In 1604 the immutable heavens of Aristotle and his orthodox disciples were again disturbed by the appearance of a new star in the constellation of Serpentarius—brighter, some declared, than the earlier star; twice as bright, said others, as Jupiter.[6] The earlier star had dawned upon an amazed and unsuspecting world which—Kepler suggested in the ironic introduction to the work which he immediately wrote—felt it to be a "secret hostile inruption," "an enemy storming a town and breaking into the market-place before the citizens are aware of his approach." This one, he declared, skilfully mingling astrol-

[6] Kepler's former teacher, Maestlin, who was among the first to observe the new star, wrote of it: "How wonderful is this new star! I am certain that I did not see it before 29th September, nor indeed, on account of several cloudy nights, had I a good view till 6th October. Now that it is on the other side of the sun, instead of surpassing Jupiter as it did, and almost rivalling Venus, it scarcely equals the Cor Leonis, and hardly surpasses Saturn. It continues, however, to shine with the same bright and strongly sparkling light, and changes colour almost every moment, now tawny, then yellow, presently purple, and red, and, when it has risen above the vapours, most frequently white" (J. J. Fahie, *Galileo, His Life and Works* [1903], p. 54).

7

ogy and irony, had come in a year and at a season of which the astrologers had predicted great things, and would therefore be acclaimed not as the secret coming of an enemy, but as the "spectacle of a public triumph, or the entry of a mighty potentate . . . then at last the trumpeters and archers and lackeys so distinguish the person of the monarch, that there is no occasion to point him out, but every one cries of his own accord—'Here we have him!' " [7] The subtle satire throughout the whole work is such that it is small wonder that the adversaries of the new astronomy sometimes hailed him as an ally, and insisted until the end that Kepler's position was that of the upholder of the orthodox astrology.

Kepler was not the only great scientist who gained both fame and infamy from his defense of new stars. On the tenth of October, 1604, Galileo Galilei, who then held the Mathematic Chair at the University of Padua, observed the *nova*. During the next few months he studied it closely, and in January, 1605, he proposed it for discussion in his public lectures—which during the preceding session had been upon the theory of the planets. No better evidence of the interest which the new star excited in the public mind can be found than the contemporary accounts of the crowds which thronged his lecture-room, forcing him to lecture in the Aula Magna of the University. The differences between his theories and those of Tycho and Kepler need not detain us here.[8] All agreed in their opposition to the accepted Aristotelian philosophy of the heavens, and

[7] Quoted in W. W. Bryant, *Kepler* (1920), p. 30.

[8] "He demonstrated that it was neither a meteor, nor yet a body existing from all time, and only now noticed, but a body which had recently appeared and would again vanish. Unlike his contemporaries, Tycho Brahe and Kepler, who thought that new stars (and comets) were temporary conglomerations of a cosmical vapour filling space; or, as is now thought, the result of some catastrophe or collision whereby immense masses of incandescent gases are produced, Galileo suggested that they might be products of terrestrial exhalations of extreme tenuity, at immense distances from the earth, and reflecting the sun's rays—an hypothesis which . . . . he also applied to comets" (Fahie, p. 55).

Kepler and Galileo both challenged the Ptolemaic astronomy by the Copernican. Galileo's public declarations on the subject mark the real beginning of the bitter antagonism which he was to encounter throughout his whole life.[9]

Even Galileo, however, had no presage, as he watched the popular excitement aroused by the discovery of a single new star, that within six years that excitement was to be multiplied a hundred fold, and that he himself, almost overnight, was to add to human knowledge not one star, as Tycho and Kepler had done, but stars innumerable, a new moon, new planets— a new world. All that the curious and observant eye of man unaided could discover was known to Galileo in 1605; but beyond the reach of human eye, shrouded as it had been since the beginning, stretched a new cosmos. Its discovery by the genius of one man might well have been called by Bacon's title, *The Greatest Birth of Time.*

II

As seven cities warred for Homer dead, so at least seven countries have warred for the honor of possessing in their annals the inventor of that simplest yet most epoch-making of instruments—the telescope. Search for its origin leads one into many paths of time and place. Was there, in any real sense, an "inventor" of the telescope, or did it develop by slow stages

[9] In answer to the attack of Antonio de Montepulciano, Galileo published in 1605 his satirical *Dialogo de Cecco di Ronchitti da Bruzene in Perpuosito del la Stella Nuova* (Padua, 1605). This is reprinted, with a modern Italian version, in the Edizione Nazionale of Galileo's works, *Le Opere di Galileo Galilei . . .* Direttore, Antonio Favaro (1891), Vol. II. In the same volume Favaro has included *Frammenti di Lezione e di Studi sulla Nuova Stella dell' Ottobre, 1604.* The controversy was continued when, after Galileo's first publication on the telescope, Baldessare Capra of Milan claimed the invention of the instrument. Galileo replied with his *Difesa contro alle Calunnie et Imposture di Baldessar Capra,* in the first part of which he defended his contentions about the new star, which Capra had also attacked. The reply, with the two papers of Capra, is reprinted in *Le Opere,* Vol. II.

from more primitive instruments whose beginnings are lost in the mist of antiquity?[10] Its possible origin is found by various modern commentators in the "merkhet" or "measuring instrument" of the Egyptians, in Arabian legend, in the "queynte mirours" and "perspectives" of Chaucer, more probably in the "glasses or diaphanous bodies" of Roger Bacon. But no record remains of discoveries made by any of these instruments. We approach the modern scientific world more certainly in the case of Thomas Digges, whose experiments, made about 1550, seem to prove that he had discovered the principle of the telescope. With one possible exception,[11] however, Digges, like his slightly later contemporary, Dr. Dee, seems to have been more concerned with magnification of objects upon earth than with observation of the celestial bodies.

So far as modern astronomy is concerned, all these early discoveries, interesting though they are, are only preliminary sketches for the finished portrait. Clearly they helped prepare the way; for it was not mere coincidence that Holland, Germany, France, and Italy all claim the invention of the telescope about 1608. Whoever the actual inventor, it is unquestionable

[10] Bibliography upon this subject is so extensive that I shall list only a few studies which I have found of particular interest or importance. Among modern treatments, the following are particularly good: A. N. Disney, C. F. Hill and Watson Baker, *Origin and Development of the Microscope* (London, 1928); C. Singer, "Steps Leading to the First Optical Apparatus" in *Studies in the History and Method of Science* (Oxford, 1921), II. 385 ff.; G. Govi, "Il Microscopio Composto Inventato da Galilei," *Atti R. Acad. Sci. Fis. Nat.*, II, Ser. 2 (Napoli, 1888). An English translation of this is given in the *Jour. Roy. Microsc. Soc.*, IX (London, 1889), 574 ff. A good general account may be found in R. T. Gunther, *Early Science in Oxford* (1923), II. 288–331. An early account of great interest is the treatment by Hieronymus Sirturus, *Telescopium sive Ars Perficiendi novum illud Galilaei Visorum Instrumentum ad Sidera . . .* (Francofurti, 1618).

[11] A modern study of Digges suggests that his concept of infinity may have been the result of the fact that he did use the telescope for astronomical observations. See Francis R. Johnson and Sanford V. Larkey, "Thomas Digges, the Copernican system, and the idea of the infinity of the universe," *Huntington Library Bulletin*, No. 5 (April, 1934), pp. 69–117.

that the lens-makers of Holland first constructed telescopes in such a way as to make them available for astronomical discovery, though no one of the early makers produced an instrument of sufficient power to make celestial observations. In 1608, within a month two spectacle-makers—Jan Lippershey of Middelburgh in Zealand and James Metius of Alkmaar—filed petitions for the exclusive right of manufacture and sale of instruments for seeing at a distance. Within the next few months a number of the instruments were manufactured and sold in various parts of Europe.

Important though it may be to the historian of science to establish the inventor of the telescope, to the historian of ideas the precise origin of an instrument is of little consequence. The lens-makers who exhibited the novelty, the aristocrats who considered it an interesting toy, might well have remained in the Paradise of Fools had it not been for one man who, recognizing instantly the potentialities of the new instrument, had the ability to develop it in such a way that it might be used for celestial observation, and the genius to interpret what he saw. No one doubts that this was Galileo Galilei. As the star of 1572 remained "Tycho's star," so the telescope almost overnight became "Galileo's tube," and such it remained to the seventeenth century. Galileo, on a visit to Venice about May, 1609, heard of the presentation by a Dutchman—probably Lippershey—to Prince Maurice of Nassau of an optical instrument which, as if by magic, brought far objects near.[12] Immediately recognizing the principle from the description, and evidently foreseeing the possibilities, Galileo set himself to the construction of an instrument with such success that he was

[12] See G. Moll, "On the First Invention of Telescopes," *Jour. Roy. Inst.* (London, 1836), pp. 319, 483, 496. Logan Pearsall Smith (*The Life and Letters of Sir Henry Wotton*, I [1907], 486 n.) finds evidence in letters of Paoli Sarpi that news of the telescope reached Venice as early as 1608, but mentions a different account given in dispatches of Giorgio Bartoli, secretary of the Tuscan resident. He speaks of a telescope being tried from the campanile of San Marco on August 22, 1609, and of another brought to Venice by a stranger a week later.

able to report to his brother-in-law on August 29, 1609, that he had made a glass which far surpassed the powers of the one reported from Flanders.[13] He describes vividly the first public test of the instrument, when after his exhibition of it before "their Highnesses, the Signoria, many of the nobles and senators, although of a great age, mounted more than once to the top of the highest church tower in Venice, in order to see sails and shipping that were so far off that it was two hours before they were seen, without my spy-glass, steering full sail into the harbour." The senators and nobles, like Galileo himself, were impressed at first only with the utility the discovery promised for naval and military operations; their recognition of its practical value was evidenced by the life-appointment of Galileo to his professorship, with an increase in salary.

The first telescope Galileo exhibited had a magnifying power of three diameters, making objects appear nine times larger; his next had a magnifying power of about eight diameters. He continued the development until within a short time he

[13] This letter has been frequently reprinted, and may be found easily available in Fahie, pp. 77–78. A very significant sentence indicates that Galileo had not actually seen any of the Dutch instruments, but, having heard of them, had readily deduced the optical principle involved. In commenting on the original report he says: "This result seemed to me so extraordinary that it set me thinking, and as it appeared to me that it depended on the laws of perspective, I reflected on the manner of constructing it, and was at length so entirely successful that I made a spyglass which far surpasses the report of the Flanders one." In the *Sidereus Nuncius (The Sidereal Messenger)*, trans. E. S. Carlos (1880), introduction, Galileo says that a report had reached his ears of a telescope constructed by a Dutchman, and, a few days later, he received confirmation of the report "in a letter written from Paris by a noble Frenchman, Jaques Badovere, which finally determined me to give myself up first to inquire into the principle of the telescope, and then to consider the means by which I might compass the invention of a similar instrument, which after a little while I succeeded in doing, through deep study of the theory of Refraction." In *Il Saggiatore* (1623), Galileo describes at greater length the logical process by which he reached his conclusion, and declares that he discovered the principle in one night, and the next day made his first instrument. He insists that the discovery was made "by the way of pure reasoning." The passage is given in Fahie, pp. 79–81.

had perfected one which, turned toward the stars, gave him some hint of the astonishment to come, and led him to devote himself feverishly, "sparing neither labour nor expense" to the development of an instrument which showed objects nearly thirty times nearer, and nearly one thousand times larger. This —the fifth telescope of Galileo—is the "optic tube" with which the astonished "Tuscan artist" viewed the night sky, piercing the heavens and opening to humanity a new heaven and a new earth. These were the discoveries which he immediately prepared to give to the world in his *Sidereus Nuncius,* the sidereal messenger which in 1610 carried abroad, to increasing excitement, news of the discoveries which were to transform human imagination.

Galileo himself mentions the proposed publication of his epoch-making work in a letter to Belisario Vinta [14] in Florence. He is, he writes, at Venice in order to arrange for the printing of what he calls "alcune osservazioni le quali col mezo di' uno mio occhiale [15] ho fatto ne i corpi celesti." He speaks in the letter of his own amazement over his observations and renders thanks to God who has made him "solo primo osservatore di cosa ammiranda et tenuta a tutti i seccoli occulta." The letter was written on January 30, 1610. It was only five months since Galileo had constructed the first instrument which he publicly exhibited; yet in that time he had not only perfected his instrument, but had already made his greatest discoveries —of the nature of the moon and of the Milky Way, most of all of the "new stars" about Jupiter. Three of these he had observed for the first time on the night of January 7, 1610; his observations of them were not, however, complete enough for

[14] The letter is given in *Le Opere,* X, 280–81.

[15] Galileo does not call his instrument *telescopio* or *telescopium* in his earliest letters and pamphlets. *Mio occhiale,* which he uses here, is a favorite of his; elsewhere he uses *perspicillum; telescopio* seems to have been first used by him in a letter on September 1, 1611. The term *telescopium* was not original with Galileo; it is ascribed by Baptista Porta to Prince Cesi, founder of the Academy of Lincei. Kepler uses the terms *conspicillum, perspicillum, specillium, pencillium.* See Edward Rosen, *The Naming of the Telescope* (New York, 1947).

publication until, after repeated observations, he had established to his satisfaction early in March the existence of four planets of Jupiter. On March 4, 1610,[16] Galileo wrote the dedication to his *Sidereus Nuncius,* on the title-page of which he offered his discoveries at one glance to an amazed generation.

Even the modern reader, accustomed to astounding scientific discoveries, feels the excitement which the seventeenth century must have exprienced in reading Galileo's account of his "incredible delight" when for the first time he observed the heavens through his "occhiale," and shares the amazement of the lonely observer in the "absolute novelty" of his discoveries. He himself suggested, in the order of their importance, the new conceptions we shall find reflected in the literature of the seventeenth century. First and most obvious was the instantaneous increase in the number of new stars, "stars in myriads, which have never been seen before," says Galileo, "and which surpass the old, previously known, stars in number more than ten times." Again, he declares: "It is a most beautiful and delightful sight to behold the body of the Moon." Beautiful and delightful indeed to Galileo; but to poets who for centuries had sung of the smooth surface, the clear even radiance of the "luminous orb," how extraordinary must have seemed his calm statement that "the Moon certainly does not possess a smooth and polished surface, but one rough and uneven, and, just like the face of the Earth itself, is everywhere full of vast protuberances, deep chasms, and sinuosities." If this new conception of the moon was not to put on end to conventional praise of the "fair Diana," it was to offer some poets a new realism in the "vast protuberances, deep chasms, and sinuosities," and, in a short space of time, to bring back into immense

[16] Fahie says (p. 86): "The book, doubtless, appeared immediately after, say, towards the middle of March." Its publication may be dated even more accurately by the letter which Sir Henry Wotton wrote to the Earl of Salisbury on March 13, 1610, in which he says that the book is "come abroad this very day." The letter is given by Smith, I, 485–87. I have discussed it in the second article in this series.

popularity, with new meaning, the old legend of a world in the moon. But these are still matters for the future. Popular as the new conception of the moon was to become, there were other early discoveries of Galileo which vied with it. For centuries poets had sung the beauties of the Milky Way, and astronomers and philosophers had disputed about its nature. In one night Galileo had solved the problem and had, as he went on to say in the *Sidereus Nuncius,* "got rid of disputes about the Galaxy or Milky Way, and made its nature clear to the very senses, not to say to the understanding." [17]

What seemed the greatest of Galileo's discoveries had been the last of them. To this in particular he calls the attention of philosophers. "That which will excite the greatest astonishment by far," he declared, "and which indeed especially moved me to call the attention of all astronomers and philosophers is this, namely, that I have discovered four planets, neither known nor observed by any one of the astronomers before my time." The effect of Galileo's discovery of supposed new planets upon the history of science has been discussed so often and so completely by astronomers to whose attention he called it that even the layman takes for granted its implications in astronomy. Yet the effect of this discovery upon poetic and religious imagination, during the century which followed, has been so slightly treated that though we recognize the theological controversies to which it gave rise, we are hardly aware of the extent to which those new planets swam into the ken of thoughtful minds, and we have missed both the "wild surmise" and the exultation of the seventeenth-century writers as

[17] In his further discussion of the Milky Way (pp. 42–43) Galileo says: "By the aid of a telescope any one may behold this in a manner which so distinctly appeals to the senses that all the disputes which have tormented philosophers through so many ages are exploded at once by the irrefragable evidence of our eyes, and we are freed from wordy disputes upon this subject, for the Galaxy is nothing else but a mass of innumerable stars planted together in clusters. Upon whatever part of it you direct the telescope straightway a vast crowd of stars presents itself to view; many of them are tolerably large and extremely bright, but the number of small ones is quite beyond determination."

they gazed with new eyes upon the new universe Galileo had unfolded to their view. We shall find it reflected not only in figures of speech drawn from the telescope itself, from the new stars, from the moon, from the Milky Way, and from the new "planets," but in new themes for literature, and the recrudescence of old themes with new meanings, most of all in a stimulation of that imagination which like

> the fleet Astronomer can bore
> And thread the spheres with his quick-darting mind.

Before we are ready to consider the effect of the *Sidereus Nuncius* in England, it will be well to see its effect upon its own generation in Italy.[18]

III

Hardly had the *Sidereus Nuncius* appeared than letters began to pour in upon Galileo. Even before the receipt of the volume in Florence, Alessandro Sertini [19] wrote on March 27,

[18] Fahie (pp. 100–101) comments upon the fact that popular excitement in Italy grew intense as the news spread. In Florence "poets chanted the discoveries and the glory of their fellow-citizen." In Venice a contemporary described the excitement as amounting to a frenzy. This account, in which Fahie is followed by most English biographers, is based to some extent upon statements in the *Telescopium* of Sirturus, and to some extent upon Favaro, who, in the national edition of Galileo's works, had brought together a mass of material in prose and poetry, chiefly Latin and Italian, which shows the popular reception of the *Sidereus Nuncius*. Much of the material in the section that follows is based upon his studies, though I have included some passages he does not quote. In addition to *Le Opere* and contemporary editions of Galileo's and Kepler's works, which contain complimentary poems, my chief sources are the following: Favaro, *Galileo Galilei e lo Studio di Padova* (Firenze, 1883); *Bibliografia Galileiana* (Rome, 1896); "Miscellanea Galileiana Inedita" in *Reale Istituto Veneto di Scienze, Lettere ed Arti, Memorie* (Venice, 1822), Vol. II; *Nuovi Studi Galileiana* (Venice, 1891); "Amici e Correspondenti di Galileo Galilei" in *Atti del R. Istituto Veneto*, Vol. LIV; Domenico Berti, "La Venuta di Galileo Galilei a Padova e la Invenzione del Telescopio," *ibid.*, Vol. XVI, Ser. iii; Sante Pieralisi, *Urbano VIII e Galileo Galilei* (Rome, 1875); Vincenzo da Filicajà, *In Morte di Vincenzo Viviana*.

[19] Favaro, *Galileo Galilei e lo Studio di Padova*, I, 390; see also *Le Opere*, X, 305–6.

1610, indicating the interest felt in Galileo's telescopic discoveries. The preceding day, he declares, upon his arrival in the Mercato Nuovo, Filippo Manelli had approached him with word that the post from Venice was bringing a package from Galileo. The news, he declares, spread in such a way that he could not defend himself from the people who wanted to know what it was, thinking it might be a telescope. When they learned that it was the new book, curiosity did not abate, especially, he adds, among men of letters. The same evening he and others had read a passage—the section dealing with the new planets—"e finalmente," he says, "è tenuta gran cosa e maravigliosa." Florence, he goes on to say, is greatly excited over telescopes, and he begs Galileo to send him one for himself. If Galileo, he suggests in a later section of the letter, wishes Andrea Salvadore to compose something about the "Medicean stars," he should write him personally; Sertini has already suggested it to Buonarroti.

Three representative letters written within the next few weeks indicate the same great interest. On April 3, 1610, Ottavio Brenzoni wrote to Galileo from Verona, expressing his gratitude for the copy of the *Sidereus Nuncius* which he had received, and saying that the least part of the work had exceeded his expectations.[20] On the same day Benedetto Castelli, writing from Brescia, speaks of the *Sidereus Nuncius* which he has already "letta e riletta più di dieci volte con somma meraviglia e dolcezza grande d'animo." [21] Two weeks later, on April 17, 1610, Frate Ilario Altobelli, writing from Ancona, says: "Il Nuncio Sidereo di V. S. Ecc^ma fa tanto strepito ch'ha potuto destarmi da un profondissimo letargo a

20 *Le Opere*, X, 309.

21 Cf. *ibid.*, pp. 310–11. Father Castelli, a Benedictine (1577–1644), one of the earliest of Galileo's astronomical disciples, was himself an important mathematician and a lifelong friend of Galileo, with whom he corresponded frequently. In October, 1613, he was called to the Mathematical Chair in the University of Pisa, from whence he wrote Galileo on November 6 that he had been forbidden to treat in his lectures of the motion of the earth, or even to hint at it. Cf. Fahie, p. 147 and *passim*.

cui soggiaccio per un lustro continuo." [22] He proceeds with a request for lenses in order that he may make observations for himself. From Naples on June 16, 1610, Orazio del Monte wrote, declaring that the invention of the telescope was a matter of the greatest satisfaction, and hailing the discovery of the new planets as equal to the discovery of a new world; their discoverer, he declares, is equal to Columbus.[23]

This was to prove a popular figure of speech with poets in Italy, as later in England. Among the complimentary verses in which the figure appears are nine Latin epigrams by one of Galileo's Scotch pupils at Padua, Thomas Seggett,[24] who was a disciple not only of Galileo, but of Kepler, and who, it is said, was the first person to send to Kepler the *Sidereus Nuncius*. Here, in more than the rhetoric of compliment, Seggett declares that Galileo makes gods of mortals by enabling them to

[22] *Le Opere*, X, 317–18; Altobelli continues: ". . . . Impazzirebbono, se fusser vivi, gli Hipparchi, i Tolomei, i Copernici, i Ticoni, e gli Egittii et i' Caldei antichi, che non hanno veduto la metà di quello che si credevano di vedere, e la gloria di V. S. Ecc^ma con sì poca fatica offusca tutta la gloria loro; del che io ne godo tanto, che niente più."

[23] Galileo was known to his contemporaries, particularly in these earlier years, as a man of letters as well as a mathematician. Among the earliest papers Galileo was invited to read before the Academy of Florence were two (1587–88) on Dante's *Inferno*. Galileo took part in the censure passed by the Accademia della Crusca on Tasso's *Gerusalemme Liberata;* his essay, *Considerazioni al Tasso*, was discovered in 1780 by the Abbe Serassi, who was collecting materials for his *Life of Tasso* (Fahie, p. 30 n.). Galileo was well known for his defense of Ariosto, whose *Orlando Furioso* he was said to have known by heart, and which he defended as vehemently as he attacked Tasso. Among actual literary works Galileo left the fragment of a play, *Capitolo in Biasimo della Toga* (1590), an outline of a comedy in prose, and several sonnets and other short poems, which Favaro has published. The most interesting of his sonnets was published as a dedicatory poem in the *Sphinx* of Malatesti. It has been reprinted by S. W. Singer in "Milton and Malatesti," *Notes and queries*, VIII, 295–96.

[24] Seggett was one of several British pupils of Galileo whom I shall mention in the second paper in this series. His "Album Amicorum," now in the Vatican Library, contains Galileo's autograph, with the date August 13, 1599. The edition of the *Sidereus Nuncius* which Kepler reprinted at Frankfort, with a preface by himself, contains Seggett's laudatory verses.

reach stars known hitherto only to the gods; that Galileo owes much to God, but that Jupiter himself owes much to Galileo; Columbus gave man lands to be conquered by bloodshed, Galileo gave man new worlds harmful to none. Which, he asks, is the greater? In one of the epigrams Seggett groups together Kepler and Galileo, introducing into his verses the *Vicisti Galilaee,* often attributed to Kepler.[25] Another interesting poem in which the comparison with Columbus is used, but which goes farther in its implications, was the complimentary poem of Johannes Faber, later published in Galileo's *Il Saggiatore.*[26] As far as the stars of heaven are distant from the earth, declares Faber, so far does Galileo outshine others; they measure tiny miles of earth or the salt tracts of the sea; Galileo climbs bright Olympus with boundless steps and eye equipped by art.

Yield, Vespucci, and let Columbus yield. Each of these
Holds, it is true, his way through the unknown sea. . . .
But you, Galileo, alone gave to the human race the sequence of stars,
New constellations of heaven.

He calls down blessings upon the tube of Galileo which brings men to the stars. Even as eyes in age see through the power of eyeglasses, he declares, so Galileo has given spectacles to the world in its failing years. He concludes:

O bold deed, to have penetrated the adamantine ramparts of
heaven with such frail aid of crystal.
Happy souls, to whom it is given to survey the citadels of
the gods through your tube, Galileo.

Chief among the poets who used the Columbus-motif in their praise of Galileo was Giambattista Marino, who devoted a section of his long narrative poem *Adone* to Galileo's dis-

[25] Favaro, *Galileo e lo Studio di Padova,* I, 399; *Le Opere,* Vol. III, Part I, pp. 188-9.

[26] *Ad Galilaeum Lynceum Florentinum Mathematicorum Saeculi Nostri Principem Mirabilium in Caelo per Telescopium Novum Naturae Oculum Inventorem.* I have given the more important sections in translation, since the original is too long to quote.

coveries. Marino introduced as one of his digressions a scene in which Mercury, explaining the wonders of the heavens to Adonis, finally speaks of the strange markings on the moon, which leads him to a long passage in praise of Galileo and his telescope. The first stanza describes the instrument itself, as future ages will know it:

> Tempo verrà, che senza impedimento
>   Queste sue note ancor fien note e chiare,
> Mercè d'un'ammirabile stromento,
>   Per cui ciò ch'è lontan vicino appare;
> E con un'occhio chiuso, e l'altro intento
>   Specolando ciascun l'orbe lunare,
> Scorciar potrà lunghissimi intervalli
> Per un picciol cannone, e due cristalli.[27]

Galileo, says Mercury, will be a "novello Endimion," again discovering the moon. Not only a new Endymion, but a new Argonaut, and another Columbus who will show man new heavens and a new earth, new light, all things new:

> Aprendo il sen de l'Ocean profondo,
>   Ma non senza periglio, e senza guerra,
> Il Ligure Argonauta al basso mondo
>   Scoprirà novo Cielo e nova terra.
> Tu del Ciel, non del mar Tifi secondo,
>   Quanto gira spiando, e quanto serra,
> Senza alcun rischio, ad ogni gente ascose
> Scoprirai nove luci, e nove cose.

In the stanzas which follow, Marino prophesies the fame of this great discoverer, who owes much to God, but to whose labors heaven itself is indebted because in them Galileo has discovered new beauties; his fame will be eternal, and the stars themselves will tell his praise with tongues of light:

> Ben dei tu molto al Ciel, che ti discopra
>   L'invention de l'organo celeste,
> Ma viè più 'l Cielo a la tua nobil'opra,

---

[27] *L'Adone, Poema del Cavalier Marino* (Venice, 1625), Canto X.

Che le bellezze sue fà manifeste.
Degna è l'imagin tua, che sia là sopra
    Tra i lumi accolta, onde si fregia e veste,
E de le tue lunette il vetro frale
Tra gli eterni zaffir resti immortale.

Non prima nò, che de le stelle istesse,
    Estingua il Cielo i luminosi rai,
Esser dee lo splendor, ch'al crin ti tesse
    Honorata corona, estinto mai.
Chiara la gloria tua vivrà con esse,
    E tu per fama in lor chiaro vivrai,
E con lingue di luce ardenti e belle
Favelleran di te sempre le stelle.

Of all the discoveries of Galileo, however, none so instantaneously appealed to imagination as that of the "planets" of Jupiter. Poets in Italy were swift to seize not only the novelty of the idea, but the opportunity offered for compliment to the Medici. Marino was typical of many when he wrote:

> E col medesmo occhial non solo in lei
>   Vedrai dapresso ogni atomo distinto,
> Ma Giove ancor sotto gli auspicii miei
>   Scorgerai d'altri lumi intorno cinto,
> Onde lassù de l'Arno i Semidei
>   Il nome lasceran sculto e dipinto.
> Che Giulio a Cosmo ceda allhor fia giusto,
> E dal Medici tuo sia vinto Augusto.[28]

Many of the verses on the subject belong merely to the literature of compliment, hailing the discoverer, offering deft flattery to the house of Medici. Others, however, were less conventional and suggest real awareness of the greatness of Galileo's discoveries. Among these is a sonnet by Piero de' Bardi, one of several poems sent to Galileo by his friend Sertini. After an octave in which De' Bardi writes of the souls of the four Medici shining in heaven, he hails Galileo:

[28] *Ibid.*, stanzas 43, 45, 46–7, 44.

Tu, Galileo, apri' l tresor de' cieli
Col vetro illustre, e i gran Toscani Regi,
Fatti stelle immortali, a noi riveli.[29]

The two most interesting and significant poetic reflections
upon the theme of the Medicean stars were those by Andrea
Salvadore and Michelangelo Buonarroti, both written by request, one as the result of Sertini's request to Buonarroti and
the other upon Galileo's own request to Salvadore.[30]

Michelangelo Buonarroti the younger, nephew of the sculptor, whose friendship for Galileo remained firm long after
public expression of interest in Galileo had ceased to be politic,
wrote in these earlier days a song of praise of various heroes
who had been immortalized as stars in heaven, none of them
greater than the Medici, honored by the new stars, and Galileo
who had honored them. He writes:

Le quattro a noi non più vedute stelle,
Ch'il linceo sguardo sol dell'alto ingegno
Tuo, Galileo, ci scuopre, albergo degno
Saranno in ciel delle quattro alme belle.[31]

The most interesting of the treatments of the Medicean stars
is *Per le Stelle Medicee* [32] of Andrea Salvadore, to whom
Galileo had written at Sertini's advice. Salvadore begins his
poem with a description of the legendary assault upon heaven
of the giants and their destruction by Jove. In the second part
he draws a parallel between the older madness and the folly
of mortals of his own time who raise their voices against heaven
and, denying the existence of the new planets, jealously refuse

[29] *Le Opere*, X, 399.

[30] Favaro comments (IX, 233–35) that Galileo himself encouraged poets
to sing of what he considered his greatest glory—the discovery of the satellites of Jupiter. Salvadore's poem was requested by Galileo, written by
Salvadore, and corrected by Galileo. It was not published until after the
death of both.

[31] *Le Opere*, X, 412, footnote to Letter 372 of August 7, 1610.

[32] *Per le Stelle Medicee Temerariamente Oppugnate, Canzone di Andrea
Salvadore*, in *Le Opere*, IX, 267–72. Favaro gives not only the text, but
(IX, 238–65) a facsimile of Salvadore's manuscript with corrections in Gali-
leo's hand.

the glory due their discoverer. As in the past, so in the future, he concludes, punishment will come upon the impious; the poem concludes:

> Inchinate tacendo i lumi ignoti,
> Ch'in ciel spiegano, alteri
> Del gran Mediceo nome, i rai lucenti;
> Lumi, ch'a Giove intorno in proprii moti
> Errando van per gl'immortal sentieri:
> Chè s'indarno tentaro
> Gli empi, che i monti alzaro,
> Il regno torgli de bei giri ardenti,
> Come co' falsi accenti
> Potrete al regno suo, lingue mendaci,
> Toglier l'eterne e luminose faci?

In addition, there were many other tributes by lesser and greater contemporaries, all suggesting the interest with which Galileo's discoveries were received.[33] One of these—a somewhat later poem—is worthy of attention less for the ideas it contains than for its author. This is a Horatian ode, the *Adulatio Perniciosa*[34] of Maffeo Barberini,[35] later Urban VIII. Obviously writing in imitation of Horace, Barberini showers com-

[33] Among these I may mention the following: Virginio Cesarini wrote a Latin poem—"eleganti e bellisimi versi latini," as Favaro calls them (*Galileo e lo Studio di Padova*, I, 403)—in honor of Galileo. Luca Valerio sent Galileo a Latin epigram (*Le Opere*, VIII, 181) in a letter from Rome on November 11, 1611, declaring that Galileo's fame will be immortal. Prince Cesi sent Galileo some verses by Demisiano (*ibid.*, p. 185); Niccolo Aggiunti also wrote in Galileo's honor (Berti, p. 2016).

[34] *Memorie Storiche del Sacerdote Sante Pieralisi* (Rome, 1875), pp. 22–24.

[35] In the early days of Galileo's difficulties with his opponents, Barberini defended him more than once. On August 28, 1629, Barberini sent Galileo the verses in which he celebrated his discoveries, saying that he proposed "to add lustre to my poetry by coupling it with your name" (quoted Fahie, pp. 185–86). In 1623 Galileo's *Il Saggiatore* appeared with a dedication to Barberini, now Pope Urban VIII. By 1624, however, Galileo was forced to realize that while Barberini had defended him, Pope Urban was adamantine in his refusal to grant even passive toleration to the new astronomy. Urban was at last persuaded that Galileo's character Simplicio in the *Dialogo* was intended as an unflattering portrait of the pope. The sequel is too well known to require comment.

pliments upon Galileo, commenting upon the importance of his discovery of the telescope, of the new knowledge of the nature of the moon and of the satellites of Jupiter, and showing particular interest in Galileo's hypothesis of the nature of sun-spots. The *Adulatio Perniciosa* is not in itself of literary interest, but it marks a chapter in the tragedy of Galileo's later life.

IV

All these writers realized the importance of Galileo's discoveries; none of them, however, caught the most significant conception which Galileo's discovery of the new planets was to bring into human thought. It was, as might be expected, a philosopher who immediately became aware of the new evidence Galileo was offering to prove the truth of an age-old belief. From his prison in Naples, Campanella—suffering for his opinions as Galileo was later to suffer—wrote on January 13, 1611, a letter [36] concerning the *Sidereus Nuncius* in which his excitement and his enthusiasm characteristically overflow. He bursts into praise of the man who has restored the true philosophy of the past, who has not only brought back the glory of Pythagoras but has given new meaning to a great biblical passage:

In astronomia nos Ptolemeus et Copernicus pudefaciebant; sed tu, Vir Clarissime, non modo restituis nobis gloriam Pythagoreorum, a Graecis subdolis subreptam, eorum dogmata resuscitando, sed totius mundi gloriam tuo splendore extinguis. "*Et vidi caelum novum et terram novam,*" ait Apostolus et Isaias: illi dixerunt, nos caecutiebamus; tu purgasti oculos hominum et novum ostendis caelum et novam terram in luna.

He hails the new heavens, the new philosophy to be born of them, mingling old tales and new proofs, fantasy and reality, and raises the question of a plurality of worlds and of the possible inhabitants of these new planets. Since the beginning of recorded history man had pondered this problem; certainly

[36] Quoted in *Le Opere*, XI, 21–26.

it has been a recurrent theme in philosophies and theologies. Now for the first time there was evidence of the senses, to put an end, so Campanella seemed to think, to human wranglings. His letters to Galileo are filled with the enthusiasm of the disciple. Galileo, he declares, is the glory of Italy; he surpasses the innovators of the past. As Amerigo gave his name to a new continent, so Galileo has given his to a new universe. Thus, as philosopher and mystic, he urges Galileo not to confine himself to mere physical science, but to go beyond his immediate predecessors, and, following the great thinkers of the past—Aristarchus, Philolaus, Pythagoras—reveal a universal philosophy.[37] Even in his early letters, written in 1611, he rejoices in the fact that Galileo's tube has at last proved conclusively to man not only that there are other planets as important as "this star our earth" but that these planets, as ours, may have their inhabitants, perhaps like ourselves or, it may be, greater than we. What remains for man is to attempt some conclusion as to the nature of these planetary dwellers.[38] Campanella prophesied more truly than he knew; the century that

[37] Cf. Edmund Gardner, *Tommaso Campanella and His Poetry* (Oxford, 1923). Campanella devoted his efforts to his attempts to force Galileo even further in his development of philosophical implications. In his *Apology* (*E. Thomae Campanellae . . . . Apologia pro Galileo . . . .* Francofurti, 1622), he discusses the charges which have been brought against Galileo. In time, however, Rome overcame reason; and in 1638, in his *Universalis Philosophia seu Metaphysica,* Campanella interpolated a clause to the effect that sentence had been passed, and the older belief therefore established. It is significant, however, that in his *City of the Sun* (ed. Morley, p. 261) he says of his dwellers in Utopia, "They praise Ptolemy. admire Copernicus, but place Aristarchus and Philolaus before him." He also indicates that his dwellers in the sun are inclined toward the idea of a plurality of worlds though they are as yet uncertain (*ibid.,* p. 263): "They are in doubt whether there are other worlds beyond ours, and account it madness to say there is nothing."

[38] Galileo himself denied the possibility of existence of inhabitants of the moon, as is clear from a letter to Giacomo Muti on February 28, 1616 (quoted by Fahie, pp. 135–36). He did not commit himself on the possibility of habitation of the planets, writing to Prince Cesi, January 25, 1613, "If the question be put to me, I will answer neither yes nor no" (Fahie, p. 134 n.).

followed, in England as in Italy, was to find in this idea a new-old theme for romance, for philosophy, for theology.

Campanella was not the only thinker of the day who seized upon the implications of the Galilean astronomy and carried them farther than Galileo himself. It is too early to enter upon the question of the stimulus Galileo's observations gave to the philosophical idea of infinity which was to re-emerge in the seventeenth century with new arguments and new significance. Although Kepler was unable to accept the idea of infinite space and infinity of worlds, which Bruno adopted even before the invention of the telescope, he found no difficulty in believing that other planets besides our own are inhabited.[39] From the beginning, with a generosity that does him credit, he accepted Galileo's conclusions with only praise for his contemporary, in spite of the fact that Galileo had disproved many of his own hypotheses. Like Campanella, he urged Galileo to carry farther the implications of his discoveries, and he himself strove to adapt his mystical philosophy to the telescopic discoveries. Immediately upon receipt of the *Sidereus Nuncius* he wrote to Galileo: "I am so far from disbelieving the four circumjovial planets, that I long for a telescope to anticipate you, if possible, in discovering two round Mars, as the proportion seems to require, six or eight round Saturn, and perhaps one each round Mercury and Venus." [40] In August,

[39] Kepler discusses the question of infinity in his *De Stella Nova in Pede Serpentarii* (1606). Burton's comment in the *Anatomy of Melancholy* (II, 63) is of interest: "Kepler (I confess) will by no means admit of Brunus' infinity of worlds or that the fixed stars should be so many Suns, with their compassing Planets, yet the said Kepler, betwixt jest and earnest . . . . seems in part to agree with this and partly to contradict." Galileo, in his *Dialogue of the World System*, argued against the dangers he felt in Bruno's conception of the infinity of space, though he felt its fascination. He wrote in a letter to F. Ingoli: "Reason and my mental powers do not enable me to conceive of either finitude or infinitude" (William Boulting, *Giordano Bruno*, p. 141).

[40] Quoted, W. Carl Rufus, "Kepler as an Astronomer," in *Johann Kepler, etc.*, p. 24. One of Galileo's letters to Kepler, written August 19, 1610, from Padua (*Le Opere*, X, 421–23) suggests the half-humorous exasperation of

1610, Kepler procured a telescope and immediately started upon his own observations, which verified Galileo's. But Kepler, even more than Galileo, was interested in the principle of the telescope. In his edition of the *Sidereus Nuncius,* Kepler took occasion not only to praise his contemporary and to make clear to his own countrymen the significance of the work, but he went farther than Galileo with the principles of optics behind the development of the instrument. As a result of his interest in these problems Kepler produced his *Dioptrice,*[41] which laid the foundation for modern telescopic instruments and for years remained the greatest authority in the field of dioptrics.

The later history of Kepler and of Galileo has been told so often that, with one exception, it need not detain us here. One minor work of Kepler's belongs to the realm of imaginative literature, and, like these earlier works of Galileo and Kepler, was to stimulate literary imagination. Kepler spent his last days in poverty and personal misery. In the last year of his life, desperate in financial need, he set himself to the composition of a philosophical romance, which contained reflections of the new science, of Galileo's discovery of the nature of the moon, of Kepler's and Campanella's ponderings upon the *incolae astrorum.* Four years after the author's death appeared the *Somnium,*[42] a lunar dream of mingled fact and fancy, a composite of old moon-legend and moon-knowledge. Here, cast

---

both of them at the reception of the *Sidereus Nuncius* by the orthodox, many of whom refused even to look through the telescope: "Quid igitur agendum? cum Democrito aut cum Heraclito standum? Volo, mi Keplere, ut rideamus insignem vulgi stultitiam. Quid dices de primariis huius Gimnasii philosophis, qui, aspidis pertinacia repleti, nunquam, licet me ultro dedita opera millies offerente, nec Planetas, nec [lunam], nec perspicillum, videre voluerunt?"

[41] *Joannis Kepleri . . . . Mathematici Dioptrice, seu Demonstratio eorum quae visui et visibilibus propter Conspicilla non ita Pridem Inventa Accidunt . . . .* (1611).

[42] *Joh. Keppleri Mathematici Olim Imperatori Somnium, seu Opus posthumus de Astronomia Lunari, Divulgatum a M. Ludovico Kepplero Filio, Medicinae Candidato* (1634).

in the form of a dream, we find one of the first modern voyages to the moon, in which new astronomical knowledge was adapted to an old theme.

So far as literary history is concerned, the *Somnium* is the last of the important works of Kepler and Galileo. In those which have been discussed were the elements of much important literature of the following century. In England as in Italy we will find the immediate response of poets and prose-writers to the *Sidereus Nuncius* and to the telescope, watch the "optick tube" becoming a novelty of the day, and see how popular interest led both to satire and to enthusiasm. We will find the figures of speech used by Galileo's contemporaries, as well as many others; will watch a new realism emerging in astronomical description, and see the older classical conceptions of the sun, moon, and stars giving way in some cases to another language. We feel the fascination of moon-literature and moon-voyages, and see how in one generation an old theme ceased to be fantasy and became serious. The influence of the telescope is to be found in both major and minor writers; if, on the one hand, it causes novel figures of speech, on the other it has much to do with the vast canvas of *Paradise Lost* and of the cosmic poems which followed. Most of all, we feel a new awareness of the expanded universe which the telescope opened to man's eyes and his imagination, and see the response of poets in characteristic ways. Dread of the magnitude of the universe, with the consequent insignificance of man, lies behind George Herbert's verses:

> Although there were some forty heavens or more,
> Sometimes I peer above them all;
> Sometimes I hardly reach a score;
>   Sometimes to hell I fall.

> O rack me not to such a vast extent;
>   These distances belong to thee:
> The world's too little for thy tent,
>   A grave too big for me.[43]

[43] "The Temper" in *The Temple* (1633).

28

At the other extreme is the exultation of such a poet as Henry More, who, having become imaginatively aware of the conception of infinite space, writes in a minor poem:

> My mightie wings high stretch'd then clapping light,
> I brush the starres and make them shine more bright.
> Then all the works of God with close embrace
> I dearly hug in my enlargèd arms.[44]

Between these two extremes are many other poets, now fascinated by the new concept of space, now drawing back, half-afraid, as

> Before their eyes in sudden view appear
> The secrets of the hoary deep.

Indeed we may feel that majestic picture in *Paradise Lost* as a symbol of the seventeenth-century attitude toward the new awareness of space which the telescope caused. As Satan stands at the massive gates, peering out into the chaos which he faces, like many another voyager in strange seas of thought, he "look'd a while Pondering his voyage." For a moment even his intrepid spirit is appalled by what he sees; but if he is symbolic in his hesitation before the new space, he is no less so in the courage—even in the curiosity—with which he makes his decision. Like the intellectual adventurers of the seventeenth century, his decision made,

> At last his sail-broad vans
> He spreads for flight, and in the surging smoke
> Uplifted spurns the ground.

[44] *Cupid's Conflict.* See *Philosophical Poems of Henry More*, ed. Geoffrey Bullough (1931), p. 113.

II. The "New Astronomy" and English Imagination

" AND new Philosophy calls all in doubt," wrote John Donne in 1611, in "The First Anniversarie," a poem marked by cosmic reflection of a sort Donne had not shown before. His "new Philosophy" was more than "new astronomy." As *Ignatius his Conclave* had implied a year earlier, accepted ideas of the Renaissance were everywhere challenged by those of the "Counter Renaissance," other political concepts cutting across conventional belief in "order" and "degree," new theories of medicine attacking the old, Copernicanism challenging accepted ideas of astronomy and astrology. Donne was right in feeling that the "new astronomy" was an important element in the melancholy of the day, leading him to feel, " 'Tis all in peeces, all cohaerence gone."

I

The "new astronomy" had a dual origin in England. Imported from Italy, it also had a native English background,[1]

[1] Since the early version of this article appeared, Francis Johnson has so thoroughly developed the background of the "new astronomy" that I

since England had produced important works of the *nova* of 1572, as well as telescopic observers who might have been better remembered had Galileo's *Sidereus Nuncius* been less dramatic than it was. When the star of 1572 appeared, John Dee published one work and wrote another, which remained in manuscript, to prove that the star was in the celestial region and therefore a *nova*. One of the most important treatments was the *Alae seu Scalae Mathematicae* of Dee's pupil, Thomas Digges, published in 1573. So significant was the treatise that Tycho Brahe in his analysis of various papers written on the phenomenon devoted the longest section of his work to an analysis of Digges' opinions. While the star of 1604 was visible in England, no work of the first importance seems to have been written there. Later English scientists were content to accept the conclusions of Galileo, Kepler and other continental writers.

So far as the telescope is concerned, there is evidence of the development of the instrument in England earlier than on the continent. John Dee worked with "perspective glasses," the advantages of which, he felt, would be their service to a commander in time of war—a belief similar to Galileo's when he developed his early telescopes. Between 1580–1590 William Bourne wrote a *Treatise on the Properties and Qualities of Glasses for Optical Purposes*. Thomas Digges carried research and invention farther. It was said that he "not onely discovered things farre off, read letters, numbred peeces of money . . . but also seven myles of declared what hath been doon at that instante in private places." American scholars have suggested that Digges' interest in the question he raises in his *Perfit Description of the Caelestiall Orbes*, "whether the world have his boundes or bee in deed infinite and without boundes," may have been the result of the fact that Digges used his telescope for celestial observation.[2]

---

omit my notes, and merely refer an interested reader to his *Astronomical Thought in Renaissance England* (Baltimore, 1937).

[2] Francis Johnson and Sanford Larkey in *Huntington Library Bulletin,*

If we must still conjecture about the celestial observations of Dee and Digges, we are on firmer ground in the case of Thomas Hariot, though final evaluation of his place in the history of astronomy must wait for the publication of his manuscripts.[3] Hariot is of particular importance in any study of the influence of astronomy on English literary imagination because of his place as scientist-royal to the "School of Night." Hariot had long been interested in perspective glasses. In his *Brief and True Report of the New Found Land of Virginia,* published in 1588, he commented on the astonishment of the natives over equipment Europeans took for granted, "as Mathematicall instruments, sea compasses, the vertue of the loadstone in drawing yron, a perspective glasse whereby was shewed many strange sights. . . ."

Hariot was corresponding with Kepler at least as early as 1606, having apparently come in contact with him through their friend John Erikson.[4] Kepler's earliest inquiries were about Hariot's theories of color and light, to which Hariot replied with a discourse on the rainbow. For two years they seem to have written largely on such problems, interchanging news of books on various phases of natural philosophy. In 1609 they were chiefly concerned with optical problems on which Kepler was engaged in connection with his *Dioptrice.* Evidently Hariot urged the study of Kepler upon his disciples, since Sir William Lower wrote him from Wales on February 6, 1610: "Kepler I read diligentlie, but therein I find what it is to be so far from you. For as himself, he hath almost put me

No. 5 (April 1934), 69–117. The point is further developed in Mr. Johnson's book.

3 While an edition of Hariot is under way, I am still dependent, as in my earlier article, upon the brief extracts in Henry Stevens, *Thomas Hariot* (London, 1900). See also F. V. Morley, "Thomas Hariot," *Scientific Monthly,* XV (1922), 60–66; Rigaud, *Account of Hariot's Astronomical Papers* (1833).

4 The early correspondence is given in *Joannis Keppleri Aliorumque Epistolae Mutuae (Frankfort, 1718);* digests of five letters are given by Stevens, *op. cit.,* pp. 178–80; Morley refers also (p. 380) to *Epistolae ad Iohannem Kepplerum* (Hanschius, 1618), which I have not seen.

out of my wits." [5] A later statement in the letter that Kepler "overthrowes the circular Astronomie" indicates that Lower was grappling with Kepler's theory that the planets move not in circles but in ellipses.

Whether Hariot's first interest in the telescope came from his foreign correspondence, or was a natural development from his own early "perspective-glasses" his papers may show. Certainly in 1610, before he had heard about Galileo's discoveries, Hariot was making important astronomical observations by means of his own instrument. Early in 1610, Hariot sent Lower a "perspective Cylinder" with instructions that Lower observe "the Mone in all his changes." Lower's comments suggest that Hariot had already made many of the same observations as Galileo on the nature of the moon. [6] Lower indeed continued to feel that it was Hariot who had been the real pioneer in the new astronomy, and urged him more than once to make public his observations and theories. Hariot, however, remained to the end little interested in persuading others; his observations were made largely for his own interest, and either through carelessness or lack of any desire for public fame, he did little to make others aware of his own importance in the history of astronomy.

No matter how interesting and important, then, were these early observations of Digges and Hariot, there is no question that the real stimulation of imagination in England among poets, philosophers, churchmen, came rather from the Italian

[5] Stevens, pp. 120–4.

[6] Letter of Feb. 6, 1610 (*British Museum Add. Mss.* 6789; quoted in Stevens, pp. 120–124): "According as you wished I have observed the Mone in all his changes. In the new I discover manifestlie the earthshine, a little before the Dichotomie, that spot which represents unto me the Man in the Moone (but without a head) is first to be seene. a little after neare the brimme of the gibbous parts towards the upper corner appeare luminous parts like starres much brighter then the rest and the whole brimme along, lookes like unto the Description of Coasts in the dutch bookes of voyages. in the full she appeares like a tarte that my Cooke made me the last Weeke. here a vaine of bright stuffe, and there of darke, and so confusedlie al over. I must confesse I can see none of this without my cylinder."

33

than from the native astronomy. Whether because, as Lower had suggested, Hariot made little effort to popularize his discoveries among any except his immediate associates, or, as is more probable, because his theories as yet lacked the cosmic significance of Galileo's, it is clear enough to one who reads the popular literature of the next few years in England that the implications which affected imagination came from Italy, and were much of a sort with the stimulation that immediately followed Galileo's pronouncements in his own country. The mind of the poet was stirred then as now by implications rather than by such careful observations of those of Hariot and Lower, and the seventeenth-century English writers were quick to read such implications into Galileo, as Campanella had done. It is the "Italian's moon" and the "Florentine's new world" we find in English poetry, not the same moon as it had appeared to Hariot in London and to Lower in Wales. To many more than Milton, the telescope remained that

> optic glass the Tuscan artist views
> At evening from the top of Fesole,
> Or in Valdarno.[7]

Indeed, English poets often read implications into Galileo that would have surprised him, as did Lovelace in his "Advice to my best Brother": [8]

> Nor be too confident, fix'd on the shore,
> For even that too borrows from the store
> Of her rich Neighbour, since now wisest know,
> (And this to Galileo's judgement ow)
> The palsie Earth it self is every jot
> As frail, inconstant, waveing as that blot
> We lay upon the Deep; That sometimes lies
> Chang'd, you would think, with's botoms properties,
> But this eternal strange Ixions wheel
> Of giddy earth, n'er whirling leaves to reel
> Till all things are inverted, till they are
> Turn'd to that Antick confus'd state they were.

[7] *Paradise Lost*, I, 288–90.
[8] *Poems of Richard Lovelace*, ed. C. H. Wilkinson (Oxford, 1930), p. 175.

Two centuries before Browning English poets felt, if they did not say:

> Lo! the moon's self,
> Here in London, yonder late in Florence,
> Shall we find her face, the thrice-transfigured.
> Curving on a sky imbrued with color,
> Drifted over Fiesole by twilight.

II

News of Galileo's discoveries travelled quickly to England, undoubtedly from many sources. Sir Henry Wotton, Ambassador to Venice, had arrived at his post on September 23, 1604, the week before the *nova* was observed by Kepler's teacher, Michael Maestlin of Tübingen, and only a little more than two weeks before Galileo first observed it. With his great interest in matters of contemporary science and the close association he maintained throughout his embassy with professors and students at Padua, he must have heard much at first hand of the popular interest in Galileo's lectures, though the letters written by him during the first year of his ambassadorship are no longer extant. Fortunately, his first letter on the telescope is still in existence, a letter written on the day the *Sidereus Nuncius* appeared in Venice, March 13, 1610. Wotton wrote to the Earl of Salisbury: [9]

Now touching the occurents of the present, I send herewith unto his Majesty the strangest piece of news (as I may justly call it) that he hath ever yet received from any part of the world; which is the annexed book (come abroad this very day) of the Mathematical Professor at Padua, who by the help of an optical instrument (which both enlargeth and approximateth the object) invented first in Flanders, and bettered by himself, hath discovered four new planets rolling about the sphere of Jupiter, besides many other unknown fixed stars; likewise, the true cause of the *Via Lactae*, so long searched; and lastly, that the moon is not spherical, but endued with many

[9] The letter, which is among the *Venetian State Papers,* was quoted by Logan Pearsall Smith, *Life and Letters of Sir Henry Wotton* (Oxford, 1907), I, 486–7.

prominences, and, which is of all the strangest, illuminated with the solar light by reflection from the body of the earth, as he seemeth to say. So as upon the whole subject he hath first overthrown all former astronomy—for we must have a new sphere to save the appearances—and next all astrology. For the virtue of these new planets must needs vary the judicial part, and why may there not yet be more? These things I have been bold thus to discourse unto your Lordship, whereof here all corners are full. And the author runneth a fortune to be either exceeding famous or exceeding ridiculous. By the next ship your Lordship shall receive from me one of the above instruments, as it is bettered by this man.

Among others who may have written back to England were a group of British students at Padua, some of them resident pupils of Galileo. Three at least were closely associated with him during the period of his spectacular discoveries, one the "Mr. Willoughby" who was the guide of that curious traveller Thomas Coryat, another the Scottish Thomas Seggett, whose epigrams on Galileo have already been mentioned. Seggett was the first student at Padua whom Wotton knew, since the ambassador's earliest official act was to secure Seggett's release from a Venetian prison into which he had been thrown for speaking too freely of a Venetian nobleman. To another Scottish student Galileo entrusted his defense against the attack of the German, Martin Horky, who, shortly after the publication of the *Sidereus Nuncius,* wrote a pamphlet against the possibility of the new planets Galileo believed he had discovered.

It is entirely possible that Kepler himself sent word of the telescopic discoveries to Hariot. Evidently Hariot had not been startled "to see every day some of your inventions taken from you." With characteristic generosity, Hariot was among the first to praise the *Sidereus Nuncius.* The letter he wrote to Lower on the subject is not yet available, if it is extant, but Lower's reply reflects the early Italian attitude: [10]

[10] The original letter is in the British Museum. It is reprinted by Stevens, pp. 116–118. Hariot does not seem to have received a copy of the *Sidereus Nuncius* at the time he wrote to Lower, since Lower in his reply says: "Send

Me thinkes my diligent Galileus hath done more in his three fold discoverie than Magellane in opening the streightes to the South sea or the dutch men that weare eaten by beares in Nova Zembla. . . . I am so affected with this newes as I wish sommer were past that I mighte observe these phenomenes also. . . . We both with wonder and delighte fell a consideringe your letter, we are here so on fire with thes thinges that I must renew my request and your promise to send mee of all sortes of thes Cylinders. . . . Send me so manie as you thinke needfull unto thes observations, and in requitall, I will send you store of observations.

As Lower and his friends in Wales were "affected with this newes," so were many readers of the *Sidereus Nuncius* in England. The widespread literary response of poets, dramatists, satirists, essayists to every aspect of the "new astronomy" is the best evidence of lay interest in science during the early years of the seventeenth century.

III

"New stars" shone in English literature throughout the century. Drummond of Hawthornden, in his "Shadow of Judgement," saw in them an indication of the end of the world:

> New stars above the eighth heaven sparkle clear,
> Mars chops with Saturn, Jove claims Mars's sphere.

"The Heavens are not only fruitful in new and unheard-of stars," wrote Sir Thomas Browne in the *Religio Medici*, "the Earth in plants and animals, but men's minds also in villany

---

me also one of Galileus bookes if anie yet be come over and you can get them." An interesting section of this letter is devoted to Kepler's *Nova Stella Serpentarii*, which Lower says he and other "Traventane Philosophers" were discussing at the moment Hariot's news of the great discoveries reached them. Among the matters under discussion was Kepler's opposition to Bruno's "opinions concerninge the immensitie of the sphaere of the starres." The letter is of particular importance in connection with the controversy of the last few years as to whether Bruno's philosophical opinions found any audience in England in the early seventeenth century.

and vices." [11] *The Anatomy of Melancholy*, particularly the "Digression of Air," is rich in conjectures of the significance of "new motions of the heavens, new stars, *palantia sidera*, comets, clouds, call them what you will." Phineas Fletcher suggested in his idyllic description of Virginia: [12]

> There every starre sheds his sweet influence
> And radiant beams; great, little, old, and new
> Their glittering rayes, and frequent confluence
> The milky way to God's high palace strew.

Cowley in his "Ode to the Royal Society" mentioned the fear inspired by the early *novae:*

> So when by various Turns of the celestial Dance,
>   In many thousand Years,
>   A Star, so long unknown, appears,
> Though Heaven it self more beauteous by it grow,
> It troubles and alarms the World below,
> Does to the Wise a Star, to Fools a Meteor show.

Samuel Butler, who lost no opportunity of poking fun at the new astronomy, described Sidrophel, gazing through his telescope at a boy flying a kite, and leaping to the conclusion that this remarkable phenomena was

> A comet, and without a beard!
> Or star, that ne'r before appeared! [13]

While new stars begin to shine in poetry immediately after 1604, they are much more common after the invention of the telescope, and figures drawn from them are customarily involved with more complex allusions to various of Galileo's discoveries.

References to the telescope are so frequent in seventeenth-century English literature that any attempt to list them would result only in a dull catalogue. The instrument became increasingly familiar to the general public. By the middle of the

[11] *Works of Sir Thomas Browne*, ed. Geoffrey Keynes (London, 1928), I, 84.
[12] *Poetical Works* (1908), I, 135.
[13] *Hudibras*, Part II, Canto III, ll. 427-8.

century it was a common spectacle in public parks, one of the holiday sports of the people. Its names are many: it is now the "Mathematicians perspicil" or the "perplexive glasse" of Ben Jonson; now the "optick magnifying Glasse" of Donne; again the "trunk-spectacle" or "trunk," the "perspective" or "the glass." Most often it is "the optick tube" or merely "tube." Its lenses are the "spectacles with which the stars" man reads "in smallest characters," as Butler said in *Hudibras*. Its appearance is frequently commented upon, as in Davenant's reference in *Gondibert* to "vast tubes, which like long cedars mounted lie." Innumerable figures were coined from it, compliments paid, insults hurled by its means. "Tell my Lady Elizabeth," the second Viscount Conway wrote of Elizabeth Cecil,[14] "that to see hir is better than any sight I can see at land, or at sea, or that Galileo with his perspective can see in heaven." James Stephen in 1615 called his mistress "my perspective glasse, through which I view the world's vanity." Marvell wrote "To the King," playing upon the spots Galileo's tube had discovered in the sun:

> So his bold Tube, Man, to the Sun apply'd,
> And Spots unknown to the bright Stars descry'd;
> Show'd they obscure him, while too near they please,
> And seem his Courtiers, are but his disease.
> Through Optick Trunk the Planet seem'd to hear,
> And hurls them off, e're since, in his Career.

The telescope could readily be adapted to religious implications, as in John Vicars' "A Prospective Glasse to Looke into Heaven," Arthur Wilson's "Faith's optic," or Donne's lines written in 1614 in connection with the "Obsequies to the Lord Harrington":

> Though God be our true glasse, through which we see
> All, since the beeing of all things is hee,
> Yet are the trunkes which doe to us derive
> Things, in proportion fit, by perspective,

14 *Conway Letters* (New Haven, 1930), p. 9.

Deeds of good men; for by their living here,
Vertues, indeed remote, seeme to be neare.

Vaughan concluded his reflection, in "They Are All Gone,"

Either disperse these mists, which blot and fill
My perspective (still) as they pass,
Or else remove me hence unto that hill
Where I shall need no glass.

The literary adaptations of the telescope were so many and various that Kepler's rhetoric in his continuation of Galileo's work seems hardly exaggerated: [15]

What now, dear reader, shall we make out of our telescope? Shall we make a Mercury's magic-wand to cross the liquid ether with, and, like Lucian, lead a colony to the uninhabited evening star, allured by the sweetness of the place? or shall we make it a Cupid's arrow, which, entering by our eyes, has pierced our inmost mind, and fired us with a love of Venus? . . . O telescope, instrument of much knowledge, more precious than any sceptre! Is not he who holds thee in his hand made king and lord of the works of God?

Throughout the century we find charming figures of speech describing "the new galaxie," the Milky Way, none lovelier than Milton's

broad and ample road, whose dust is gold,
And pavement stars, as stars to thee appear
Seen in the galaxy, that milky way
Which nightly as a circling zone thou seest
Powdered with stars.

The literature of Galileo's moon is so extensive that one might devote a long section to that "coast i' th' Noone (the Floren-tine's new World)" as Phineas Fletcher called it in "The Locusts." The new "planets" became involved with the new world in the moon, as half-curiously, half-fearfully man began to question whether moon and planets are inhabited. Drum-

[15] Preface to *Dioptrice*, trans. E. S. Carlos (Oxford-Cambridge, 1880), pp. 86, 103.

mond saw in the discovery of new planets as of new stars still another omen of the end of the world: [16]

> New worlds seen, shine
> With other suns and moons, false stars decline,
> And dive in seas; red comets warm the air,
> And blaze, as other worlds were judged there.

To Dryden, later in the century, the possibility of such worlds bespoke the goodness of parental Deity: [17]

> Perhaps a thousand other worlds that lie
> Remote from us, and latent in the sky,
> Are lightened by his beams, and kindly nurs'd.

Optimism and pessimism, we shall find, were both involved in the "new astronomy." From the extended cosmos discovered by the telescope, one group might turn, as did Pope, declaring: [18]

> Thro' worlds unnumber'd tho' the God be known,
> 'Tis ours to trace him only in our own.

Others believing that "boundless mind affects a boundless space," exulted with Young: [19]

> The soul of man was made to walk the skies,
> Delightful outlet of her prison here!
> There, disencumber'd from her chains, the ties
> Of toys terrestrial, she can rove at large,
> There, freely can respire, dilate, extend,
> In full proportion let loose all her powers;
> And, undeluded, grasp at something great.

IV

The Elizabethan dramatist whose imagination would have responded most sensitively to the poetic implications of the "new astronomy" died too early to know them. Had he lived

16 *The Shadow of the Judgement,* 1630, in *Poems of Drummond,* ed. W. C. Ward, II, 61.

17 *Eleanora,* II, 76–9.   18 *Essay on Man,* I, 21–2.

19 *Night Thoughts,* Ninth Night, ll. 1061–66.

longer, Christopher Marlowe might well today pre-empt the place accorded by literary students to John Donne, as the first English poet whose imagination was stirred by the new discoveries. And I venture to suggest—since one may comfortably surmise about the dead—that the "new Philosophy" would not have called all in doubt to Marlowe. Optimism rather than pessimism, exultation rather than despondency might have been the early note of the "new astronomy" in England, if Miss Spurgeon is correct in her analysis of Marlowe's dominant images. She says: [20]

Indeed this imaginative preoccupation with the dazzling heights and vast spaces of the universe is, together with a magnificent surging upward thrust and aspiration, the dominating note of Marlowe's mind. He seems more familiar with the starry courts of heaven than with the green fields of earth, and he loves rather to watch the movements of meteors and planets than to study the faces of men. No matter what he is describing, the pictures he draws tend to partake of this celestial and magnificent quality.

Certainly the creator of Doctor Faustus would have followed the "new science," only beginning to dawn when he died, and the creator of Tamburlaine might have sent his insatiable conqueror to ride in triumph through other worlds than ours. Had Marlowe lived to ripeness, he might well have set a different pattern than did Donne for the literary reception in England of the "new Philosophy."

Because of his premature death Marlowe remains an Elizabethan while Donne has become a "modern." Their contemporary Shakespeare, who lived on into the modern world of new stars and telescopes, was an Elizabethan, so far as response to new scientific theories was concerned. Shakespeare must have seen the new star of 1604, must have heard of Galileo's discoveries in 1610. He was writing some of his greatest plays during the period that saw John Donne's transition from "Elizabethan" to "modern." Yet his poetic imagination showed no response either to new stars or to other

[20] Caroline Spurgeon, *Shakespeare's Imagery* (New York, 1935), p. 13.

spectacular changes in the cosmic universe. If Shakespeare knew any of the theories of Bruno or of the "School of Night," he saw in them only a chance for satire in *Love's Labour's Lost*. So far as I can see, they left no lasting imprint upon his imagination.[21] He showed no such interest as did Bacon in the *primum mobile*. His most recurrent astronomical figure is that the stars and planets keep their motion in their spheres. His imagination was not stirred by concepts far removed from man's experience. Comparing Shakespeare's range with Marlowe's, Miss Spurgeon rightly says: "With Shakespeare, it is far otherwise. His feet are firmly set upon 'this goodly frame, the earth,' his eyes are focussed on the daily life around him."

If Shakespeare ever expressed himself on the new cosmology, he should have done so in *King Lear*, written while men's minds were dwelling on the significance of the new star of 1604. He might have used the star of 1604 as the most spectacular of all those "dire portents" in *Lear*, the only play in which he paid much attention to cosmology. Dire portents, troubled heavens, the influence of planets and stars on "order" in the little world—all these analogies sprang to his mind, in connection with "late eclipses" in the sun and moon. The eclipses were real enough: a nearly total eclipse of the sun on October 2, 1605, a partial eclipse of the moon on September 27 of the same year. Yet both Shakespeare and his audience must have

[21] Miss Spurgeon has tried to prove that Shakespeare was aware of the Copernican hypothesis. In *Troilus and Cressida*, when Ulysses discusses order in the heavens, she thinks that Shakespeare refers to the Copernican theory when Ulysses says that "the glorious planet Sol" is "in noble eminence enthroned and sphere Amidst the other." Miss Spurgeon has read the lines out of context. Earlier in his speech Ulysses had said,

> The heavens themselves, the planets, *and this centre,*
> Observe degree, priority, and place.

To Shakespeare, as to Ulysses, Earth still remained "this centre." Sol was sphered not in the midst of a Copernican system but merely in a central place among the Ptolemaic planets—of which our earth was not yet one. Three planets, Saturn, Jupiter, Mars, were above the sun; three, Venus, Mercury, and Luna, were below the sun. Only in this sense, long accepted by the orthodox, was the sun a centre to Shakespeare.

been aware that more ominous than those had been the appearance of a new star in a heavenly constellation. Perhaps he felt that it would be too obvious an anachronism to read back into the period of Lear a phenomenon associated by his audience with the immediate present. Perhaps his failure to refer to the *nova* arose from the fact that he was never so interested in topical references as was his contemporary Ben Jonson, who was one of the earliest English dramatists to refer to Galileo's telescopic discoveries.

There was probably nothing significant in *The Speeches at Prince Henry's Barriers*.[22] Although it is possible to read into Jonson's star-references awareness of the discovery of new stars, his uses of astronomical and astrological matters in the masque are still conventional in nature. But in the short space of time that intervened between the *Barriers* and *Love Freed from Folly and Ignorance,* Jonson, like many other Englishmen, had become acquainted with the new ideas. In the cryptic dialogue between *Love* and *Sphynx* he referred to Galileo's discoveries, within a few months after the news reached England.

> SPHYNX: I say: you first must cast about
> To find a world the world without.
> LOVE: I say, that is already done,
> And is the new world in the moon.
> SPHYNX: Cupid, you do cast too far;
> This world is nearer by a star:
> So much light I give you to't.
> LOVE: Without a glass? well, I shall do't.

From this time on, allusions to various aspects of the new astronomy are fairly common in Ben Jonson. The most extended occur in *News from the New World Discovered in the*

22 *The Speeches at Prince Henry's Barriers,* according to Hereford and Simpson, was presented January 6, 1610; Henry Morley had given the date as January 1, 1611. *Love Freed from Folly and Ignorance* was presented February 3, 1611. *News from the New World* was presented ten years later on January 6, 1621. *The Staple of News* was performed soon after the coronation on February 2, 1626.

Moon and in *The Staple of News*. When the Heralds, pretending a new method of collecting news, suggest that their information has come from the Moon "by moon-shine," the Printer, not to be outdone in familiarity with telescopic discoveries, declares:

Oh, by a trunk! I know it, a thing no bigger than a flutecase: a neighbor of mine, a spectacle-maker, has drawn the moon through it at the bore of a whistle, and made it as great as a drum-head twenty times, and brought it within the length of this room to me, I know not how often.

But the Chronicler, equally versed in all that is sophisticated in the way of news, declares that this is no longer new. "Your perplexive glasses," he says, "are common." As the dialogue continues, each character interrupting the others in order to be the first to report current gossip, we hear "Of a new world . . . And new creatures in that world . . . In the orb of the moon, which is now found to be an earth inhabited . . . with navigable seas and rivers . . . Variety of nations, policies, laws . . ." There is even humorous discussion of a question which seventeenth-century wits liked to raise, "whether there are inns and taverns there." Finally with malicious wit, Jonson suggests that "the bretheren of the Rosie Cross have their college within a mile of the moon; a castle in the air." As the climax to the masque, "the Scene opes and discovers the Region of the Moon." The masque seems to have hit the popular fancy, and to have been successful to a greater degree than its successor *The Staple of News,* in which Jonson repeated the telescopic theme and elaborated his satire of news-gathering to a tedious extent.

Only one passage of *The Staple of News* need concern us here. When Pennyboy urges Cymbal to let the princess hear some news,

> Any, any kind
> So it be news, the newest that thou hast,
> Some news of state for a princess,

Thomas, second clerk of the office, reads that the King of Spain has been chosen Pope, and Spinola made general of the Jesuits, at which Fitton comments:

> Witness the engine that they have presented him,
> To wind himself with up into the moon
> And thence make all his discoveries!

Later in the same scene he reads in the news from Florence:

> They write was found in Galileo's study
> A burning-glass, which they have sent him too,
> To fire any fleet that's out at sea.

Ben Jonson unconsciously prophesies for the century recurrent fictional devices used in the writing of the many fantasies whose common theme was the attempt of mortals to reach the moon. "There are in all," declares his Herald,[23] "but three ways of going thither: one is Endymion's way, by rapture in sleep, or a dream. The other Menippus's way, by wing, which the poet took. The third, old Empedocles's way; who, when he leaped into Aetna, having a dry sear body, and light, the smoke took him, and whift him up into the moon." "Empedocles's way," we do not find; but the ways of Endymion and of Menippus—the one by means of a dream, the other by wing—were persistent, though through the development of science they suffered a sea-change. One and all seventeenth-century voyagers to the moon "flew," as Jonson prophesied, "upon the wings of his muse."

V

Among English poets, none showed a more immediate response to the new discoveries than John Donne, nor is there a more remarkable example of the effect of the *Sidereus Nuncius*. Whether there was depth in the manifold interests of Donne some critics question; no one who observes his reaction to novelty can doubt the breadth of his interests. His curious mind—that "hydroptic, immoderate thirst of human languages

23 *News from the New World,* p. 358.

and learning"—was always hungry for new fare; for a time he fastened upon the new astronomy as another source for figures of speech, another vehicle for his restless imagination. No matter what our ultimate conclusion may be as to the effect upon him of the Copernican or the Galilean hypotheses, no one who reads his poetry thoughtfully, with due attention to chronology, can fail to see the stimulation of his mind from those two great moments in the history of astronomy—the discovery of Kepler's new star of 1604 and Galileo's contagious enthusiasm in the *Sidereus Nuncius.*

The *Songs and Sonets,* the majority of which were written before the turn of the century, contain no significant astronomical figures of speech. References to the sun, moon, and stars appear, though not in the proportion we find later, but these are purely conventional:

> And yet no greater, but more eminent,
>> Love by the Spring is growne;
>> As, in the firmament,
> Starres by the Sunne are not inlarg'd, but showne.[24]

The same conventions appear in the *Epigrams,* the *Elegies and Heroicall Epistles,* and the *Satyres,* in which the figures of speech drawn from the heavens are even fewer than those in the *Songs and Sonets,* and equally lacking in significance.[25] Even in the first *Progresse of the Soule,* in which one might expect some stirring of the cosmic imagination, we do not find it. Not unjustly we may conclude that Donne unconsciously suggested his own early attitude in "Elegie XVIII":

[24] *Loves Growth* in *Complete Poetry and Selected Prose,* ed. John Hayward (Bloomsbury, 1929), p. 24. Cf. *Confined Love,* p. 26; *A Valediction: of Weeping,* p. 28; *A Nocturnall upon S. Lucies Day,* p. 32.

[25] The only one of the *Songs and Sonets* in which a phrase suggests the astronomical figures of speech peculiar to later poems is *The Primrose,* pp. 45–6. Here Donne compares the "Primrose hill" to the terrestrial galaxy:

> And where their forme, and their infinitie
>> Make a terestriall Galaxie,
> As the small starres doe in the skie.

Grierson, *Poems of John Donne* (Oxford, 1912), II, 8 ff., places *The Prim-*

> Although we see Celestial bodies move
> Above the earth, the earth we Till and love.[26]

In the *Verse Letters to Severall Personages* we begin to feel the stirring of a different kind of imagination. In one of the poems to the Countess of Bedford, we find Donne's first poetic references to the Copernican theory and his association of the "new astronomy" with the "new Philosophy":

> As new Philosophy arrests the Sunne
> And bids the passive earth about it runne,
> So we have dull'd our minde.

At this time new stars begin to shine in his poetry. The stars of 1572 and 1604 came to his mind when he sought a figure of speech to describe the geographical and astronomical expansion of his own period:

> We have added to the world Virginia, and sent
> Two new starres lately to the firmament.[27]

Both the new star of 1604 and the comet of 1607 came to mind when he wrote to the Countess of Huntingdon:

> Who vagrant transitory Comets sees,
> Wonders, because they are rare: But a new starre
> Whose motion with the firmament agrees,
> Is miracle, for there no new things are.[28]

The regret which Tycho and Kepler, in pre-telescopic days, may have felt as they observed the disappearance of new stars is caught by Donne in three passages:

> But, as when heaven lookes on us with new eyes,
> Those new starres every Artist exercise,

---

rose in the third group, some of which, he thinks, were written after Donne's marriage, and suggests a date as late as 1612 for *Twicknam Garden*. He does not date *The Primrose* (p. 48). The galaxy reference suggests that the poem was written after 1610.

[26] *Elegie*, XVIII, p. 94.

[27] *To the Countesse of Bedford*, pp. 165–6.

[28] *To the Countesse of Huntingdon*, p. 169.

What place they should assigne to them they doubt,
Argue, and agree not, till those starres goe out.

At another time he looked back to the Golden Age when man's
leisurely life was so indefinite that

> if a slow pac'd starre had stolne away
> From the observers marking, he might stay
> Two or three hundred yeares to see't againe,
> And then make up his observation plaine.

Again he wrote, unconsciously predicting his own later attitude
toward such figures of speech:

> And in these Constellations then arise
> New starres, and old doe vanish from our eyes.[29]

For a short time new stars eclipsed the old in Donne's mind,
then vanished from his poetry, either because the possibility
of *novae* became so widely accepted that the metaphor ceased
to be novel or because Donne's imagination passed on to other
stimuli. But the new stars did not really become important
to Donne until they were interpreted in the light of Galileo's
discoveries.

The *Sidereus Nuncius,* we remember, appeared in Venice on
March 13, 1610. Donne's prose satire, the *Conclave Ignatii,*
was entered in the *Stationer's Register* on January 24, 1611;
on May 18 of the same year an English translation was entered
under the title, *Ignatius his Conclave.* The work began in
Donne's mind as a Lucianic "Dialogue in Hell." He introduced "innovators" of the Renaissance, in "a secret place,
where there were not many, beside Lucifer himselfe; to which,
onely they had title, which had so attempted any innovation
in this life, that they gave affront to all antiquities, and in-
duced doubts, and anxieties, and scruples, and after, a libertie
of beleeving what they would; at length established opinions,
directly contrary to all established before." Here he found
Machiavelli, Paracelsus, Ignatius, and, most important for our
purposes, Copernicus:

29 *Funerall Elegie,* p. 211; *First Anniversary,* pp. 200, 204.

As soone as the doore creekt, I spied a certaine Mathematitian, which till then had bene busied to finde, to deride, to detrude Ptolomey; and now with an erect countenance, and setled pace, came to the gates, and with hands and feet (scarce respecting Lucifer himselfe) beat the dores, and cried: "Are those shut against me, to whom all the Heavens were ever open, who was a Soule to the Earth, and gave it motion?"

By this I knew it was Copernicus . . . To whom Lucifer sayd: "Who are you?" . . . "1 am he, which pitying thee who wert thrust into the Center of the world, raysed both thee, and thy prison, the Earth, up into the Heavens; so as by my meanes God doth not enjoy his revenge upon thee. The Sunne, which was an officious spy, and a betrayer of faults, and so thine enemy, I have appointed to go into the lowest part of the world. Shall these gates be open to such as have innovated in small matters? and shall they be shut against me, who have turned the whole frame of the world, and am thereby almost a new Creator?"

Although *Ignatius his Conclave* took form in Donne's mind as a dialogue in Hell, as published it seems to begin as another literary form, also ultimately derived from Lucian, a cosmic voyage: "I was in an Extasie," wrote the author, "and

> My little wandring sportfull Soule
> Ghest, and Companion of my body,

had liberty to wander through all places, and to survey and reckon all the roomes and all the volumes of the heavens, and to comprehend the situation, the dimensions, the nature, the people, and the policy, both of the swimming Ilands, the Planets, and of all those which are fixed in the firmament." Only at the beginning and toward the end of the work does Donne do anything with his cosmic voyage, with the result that *Ignatius* is a curious mingling of two literary *genres*. In the original Dialogue of the Dead Copernicus is a major character; at the beginning and end of the work Donne is interested in Galileo and Kepler, who are mentioned immediately after the passage on "the swimming Ilands, the Planets, and of all that are fixed in the firmament." "Of which," Donne writes, "I thinke it an honester part as yet to be silent, than to do

Galileo wrong by speaking of it, who of late hath summoned the other world, the Stars to come nearer to him, and give him an account of themselves. Or to Keppler, who (as himselfe testifies of himselfe) *ever since Tycho Braches death hath received it into his care, that no new thing should be done in heaven without his knowledge."*

Later in the dialogue Donne returns to the theme of Galileo's discoveries, when Lucifer trying "earnestly to thinke, how he might leave Ignatius out" decided on a device:

I will write to the Bishop of Rome: he shall call Galilaeo the Florentine to him; who by this time hath thoroughly instructed himselfe of all the hills, woods, and Cities in the new world, the Moone. And since he effected so much with his first Glasses, that he saw the Moone, in so neere a distance that hee gave himselfe satisfaction of all, and the least parts in her, when now being growne to more perfection in his Art, he shall have made new Glasses, and they received a hallowing from the Pope, he may draw the Moone, like a boate floating upon the water, as neere the earth as he will. And thither . . . shall all the Jesuites bee transferred. . . . And with the same ease as you passe from the earth to the Moone, you may passe from the Moone to the other starrs, which are also thought to be worlds, and so you may beget and propagate many Hells, and enlarge your Empire.

In *Ignatius* "Jack" Donne was smiling over a "new astronomy" that did not yet matter to him. Within a year he began to understand its implications more fully. Early in 1610 occurred the death of Elizabeth Drury, daughter of Sir Robert Drury, who was to be Donne's patron. At the time of the young girl's death, Donne wrote his *Funerall Elegie,* conventional enough in its references. For the anniversary of the death in 1611 he composed *The First Anniversarie.* Barely a year had passed, yet Donne who had laughed in *Ignatius* had come to realize that "new Philosophy" might indeed call all in doubt.

The themes of the *Anniversary Poems* were literary commonplaces. Donne himself had used most of them before, though never with such moroseness. Elizabeth Drury was not the real subject of Donne's poem. The death of a young girl was only his point of departure.[30] When we turn from the *Funerall Elegie* to the poem written only a year later on the anniversary of Elizabeth Drury's death, we find a changed world and a man aware of change. All the former motifs are here; the "Decay of Nature" is the central theme; the new stars appear and fade again, but they have ceased to be mere figures of speech, and have taken on new meaning, as Donne sees the relation to cosmic philosophy. They are a symbol of the "Disproportion" and the "Mutability" in the universe of which Donne has become compellingly aware: [31]

> It teares
> The Firmament in eight and forty sheires,
> And in these Constellations then arise
> New starres, and old doe vanish from our eyes.
> As though heav'n suffered earthquakes, peace or war,
> When new Towers rise, and old demolish't are.

It is man, he is forced to conclude, who has slain "Proportion": [32]

> Man hath weav'd out a net, and this net throwne
> Upon the Heavens, and now they are his owne.
> Loth to goe up the hill, or labour thus
> To goe to heaven, we make heaven come to us.
> We spur, we reine the starres, and in their race
> They're diversly content t'obey our pace.

Donne's most quoted lines, in which he reflects the poignant regret of a generation which had inherited from the past centuries conceptions of *order, proportion, unity,* which had felt the assurance of the immutable heavens of Aristotle, take on new meaning when one reads them, remembering the revolution in thought that was occurring in 1610: [33]

[30] I have developed this idea at length in *The Breaking of the Circle* (Evanston, 1950).

[31] *First Anniversarie,* ll. 257–262.

[32] *Ibid.,* ll. 279–284.

[33] *Ibid.,* ll. 205–214. In the light of Jonson's indictment of this poem, recorded by Drummond of Hawthornden, it is interesting to notice how

And new Philosophy calls all in doubt,
The Element of fire is quite put out;
The Sun is lost, and th' earth, and no mans wit
Can well direct him where to looke for it.
And freely men confesse that this world's spent,
When in the Planets, and the Firmament
They seeke so many new; then see that this
Is crumbled out againe to his Atomies.
'Tis all in peeces, all cohaerence gone;
All just supply, and all Relation.

Here is Copernicanism to be sure, but it is less the position of this world than the awareness of new worlds that troubles the poet; less the disruption of this little world of man than the realization how slight a part that world plays in an enlarged and enlarging universe that leads Donne to his conclusion.

The *First Anniversarie* marks the climax of Donne's interest in the Galilean astronomy. Indeed, were Donne the main interest of this essay, it would be tempting to study in detail a highly significant trait of his imagination, admirably illustrated by his reaction to the "new stars" and Galileo's discoveries: his almost immediate response to new ideas, followed by cooling of his interest. The "new star" motif, as we have seen, is persistent from his first use of it in the *Verse Letters*

---

this particular passage reechoed in Drummond's mind some years later when he wrote his *Cypress Grove* (1623). In the following passage, whole phrases are taken from Donne (*A Cypress Grove*, ed. Samuel Clegg [1919], p. 35): "The element of fire is quite put out, the air is but water rarefied, the earth is found to move, and is no more the centre of the universe, is turned into a magnet; stars are not fixed, but swim in the ethereal spaces, comets are mounted above the planets. Some affirm there is another world of men and sensitive creatures, with cities and palaces, in the moon; the sun is lost, for it is but a light made of the conjunction of many shining bodies together, a cleft in the lower heavens, through which the rays of the highest diffuse themselves; is observed to have spots. Thus sciences, by the diverse motions of this globe of the brain of man, are become opinions, nay, errors, and leave the imagination in a thousand labyrinths. What is all we know, compared with what we know not?"

through those letters—that is, it appears some time between 1604 and 1609 and persists until about 1614. After that it almost disappears. The *Sidereus Nuncius* obviously affected him greatly at the time; those themes, too, tend to disappear. This was, of course, partly the result of the change in Donne's own personal life: his visit to the continent with Sir Robert Drury, followed by his entrance into Holy Orders, the death of his wife—all these turned his thoughts in other directions. One must remember that immediately after the period dealt with here occurred the change from his writing on secular subjects to the period of his religious poetry and prose. While Donne continued to refer to astronomical ideas in his letters and in such personal works as his *Devotions upon Emergent Occasions,* references to the new astronomy—both Copernican and Galilean—are rare in his formal *Sermons* and *Divine Poems.* He occasionally mentioned the telescope, as when he wrote to Mr. Tilman: [34]

> If then th'Astronomers, whereas they spie
> A new-found Starre, their Opticks magnifie,
> How brave are those, who with their Engine, can
> Bring man to heaven, and heaven againe to man?

or when he says, in a sermon: [35]

God's perspective glass, his spectacle, is the whole world . . . and through that spectacle the faults of princes, in God's eye, are multiplied far above those of private men.

In another sermon we find a reference to Galileo's discovery of the nature of the Milky Way, which recalls the earlier figure in *The Primrose:* [36]

In that glistering circle in the firmament, which we call the Galaxie. the milkie way, there is not one starre of any of the six great magnitudes, which Astronomers proceed upon, belonging to that circle: it

[34] "To Mr. Tilman After He Had Taken Orders," *ed. cit.,* p. 305.
[35] *Sermon Preached in the Evening of Christmas-Day,* 1624, *The Works of John Donne,* ed. Henry Alford (London, 1839), I, 26.
[36] *Donne's Sermons,* ed. Logan Pearsall Smith (London, 1920), p. 221.

is a glorious circle, and possesseth a great part of heaven, and yet it is all of so little starres, as have no name, no knowledge taken of them.

Only occasionally in the sermons does Donne venture upon more philosophical connotations of the Galilean discoveries. The idea of a plurality of worlds, for a churchman, was indeed a dangerous tenet, even, as it came to be called, the "new heresy." The condemnation of Bruno listed that belief as one of the chief charges against him; many orthodox Protestants, as well as Catholics, felt that such a conception struck at the roots of the Christian idea of the sacrifice of Christ. In his *Holy Sonnets,* written after the death of his wife in 1617, Donne does not hesitate to suggest the possible existence of other worlds, though without theological connotation: [37]

> You which beyond that heaven which was most high
> Have found new sphears, and of new lands can write,
> Powre new seas in mine eyes, that so I might
> Drowne my world with my weeping earnestly.

In his *Devotions upon Emergent Occasions,* he plays with the idea: [38]

Men that inhere upon Nature only, are so far from thinking, that there is anything singular in this world, as that they will scarce thinke, that this world it selfe is singular, but that every Planet, and every Starre, is another World like this; They find reason to conceive, not onely a pluralitie in every Species in the world, but a pluralitie of worlds.

But he is cautious in his expression upon one of the few occasions in which he raised the problem in his sermons: [39]

And then that heaven, which spreads so farre, as that subtill men have, with some appearance of probabilities, imagined, that in that heaven, in those manifold Spheres of the Planets and the Starres,

[37] Sonnet V, *ed. cit.,* p. 281.
[38] *Devotions upon Emergent Occasions,* ed. Hayward, p. 514.
[39] *Sermons,* ed. Logan Pearsall Smith, p. 352. The date of the sermon is

there are many earths, many worlds, as big as this world which we inhabit.

Only on one occasion does he approach the vexing question of orthodox theology, and then, so guarded is his statement that it is difficult to tell whether he is really referring to the "new heresy." He writes in one of his more dramatic sermons of [40]

the merit and passion of Christ Jesus, sufficient to save millions of worlds, and yet, many millions in this world (all the heathen excluded from any interest therein) when God hath a kingdom so large, as that nothing limits it . . .

Whether the churchman found it expedient in his sermons to keep away from those disputed matters, or whether the poet had ceased to feel the appeal of figures of speech that once had led him to new reaches of poetry is a matter that cannot be determined.

Ideas that were startling and revolutionary in 1610 were becoming more familiar. But the "new Philosophy" did not cease to be "new" in Donne's lifetime, did not settle into a literary convention that was almost a commonplace as it became a half-century later. The unrest Donne had felt in 1611 at the disruption of his universe was still felt by Pascal with his terror of the silence of infinite spaces. Donne apparently ceased to ponder the new hypotheses. Perhaps his attitude was that of Milton's Angel, who recognized the appeal of the new ideas to man's curiosity, but warned Adam:

> Solicit not thy thoughts with matters hid;
> Leave them to God above; him serve and fear.

1629. Another reference to the same idea is to be found in a sermon of 1624 (ed. cit., p. 160):

"Let every Starre in the firmament, be (so some take them to be) a severall world, was all this enough?"

[40] *Works,* ed. Alford, VI, 6–7. It will be noticed that most of these references are fairly early in Donne's religious writing. They gradually disappear; the few that I have found later are of comparatively little significance.

"Paradox and Probleme" Donne remains to his modern critics, who will probably never agree about the "conversion" that transformed Jack Donne into Dr. John Donne, Dean of St. Paul's. Read against the scientific background of his time, against the inruption of new stars and the dramatic discoveries of Galileo's telescope, his experience seems an epitome of the experience of his generation. I shall continue to believe that the discoveries of the new astronomy, coinciding with a troubled period in his own personal life and in his age, proved the straw that broke the back of his youthful scepticism and led John Donne "from the mistresse of my youth, Poesy, to the wife of mine age, Divinity."

III. *Kepler, the Somnium,*
and John Donne

IT IS one of the ironies of history that the *Somnium* [1] of Johann Kepler should have been almost completely neglected by historians both of science and of literature. Yet in its final form, it was the last work of a great scientist; its notes include Kepler's final pronouncements on matters of great importance in both physics and astronomy. As a work of literature, it is important as the first modern scientific moon-voyage, and a source of many of the later "cosmic voyages" of the seventeenth and eighteenth centuries. It is also of unusual biographical significance, since it throws light upon certain obscure matters in Kepler's life. In addition, the *Somnium* had an immediate

---

[1] *Joh. Keppleri Mathematici Olim Imperatorii Somnium sive Opus posthumum de Astronomia Lunari. Divulgatum a M. Ludovico Kepplero Filio, Medicinae Candidato. . . .* (Francofurti, 1634). The text may be found also in *Joannis Kepleri Astronomi Opera Omnia edidit* Dr. Ch. Frisch (Francofurti, 1870), Vol. VIII. No English translation of the *Somnium* has been published. Since this article appeared, a complete translation of both text and notes has been made by one of my graduate students, Joseph Lane. The typescript is deposited in the Carpenter Library, Columbia University.

effect upon English astronomy, and also a curious effect upon at least one important English poet.

I

In the form in which it was posthumously published in 1634, the *Somnium* included, in addition to the brief tale from which the volume takes its name, a long series of *"notae in Somnium Astronomicum,"* an appendix containing notes upon these notes, the first Latin translation of Plutarch's *De Orbe in Facie Lunae,* and a series of notes upon that work. I am concerned, however, only with the original tale.

Like so many imaginary voyages, before and after, the *Somnium* is cast into the form of a dream. Kepler relates that in the year 1608, when discord was raging between the brothers Prince Rudolf and Archduke Matthias, the author became interested in reading Bohemian legends, particularly those concerning the Libyan virago, most celebrated in the art of magic. One night, after a period of reading and of contemplation of the heavenly bodies, he fell into a deep slumber, and seemed to be reading another book of which he tells the general intent.

The tale within a tale has to do with the fortunes of a young man named Duracotus, a native of Islandia, "which the ancients call Thule." He was the son of remarkable parents; his father he did not remember, but, according to the account given him by his mother, he had been a fisherman, who died at the ripe age of 150 years, when his son was still an infant. The mother, Fiolxhilda, was a "wise woman," who supported herself by selling mariners little bags of herbs which contained mysterious charms. Unfortunately Fiolxhilda was a woman of ungovernable temper; upon one occasion when her young son curiously examined the contents of one of the bags, she became inflamed with anger, and impulsively pledged the boy to the captain in place of the little sack he had destroyed, in order that she might retain the money.

For a time the mother disappears from the tale, and we follow the fortunes of the son, whom Kepler portrays with sym-

59

pathy. We accompany him on a voyage between Norway and England, and see him arrive in Denmark. Violently ill from the rough sea, he is of little use to the captain, who is glad to rid himself of an incubus. Since the captain is carrying letters to the Danish astronomer, Tycho Brahe, on the Island of Wena, he dismisses the boy as messenger, promising to return for him in time. For some years Duracotus remained with the great astronomer, who saw such promise in him that, when the captain returned, Tycho refused to send the youth home. So he remained, learning the lore of astronomy, "the most divine of sciences." He was particularly interested in the fact that Tycho and his students "studied the stars and moon for whole nights with wonderful machines, a fact which reminded me of my mother, since she was also accustomed to hold assiduous colloquy with the moon."

After five years Duracotus returned home, happy to find his mother still alive, often repenting the fit of temper in which she had sent her son away. To his surprise, he came to realize that the "wise woman" was as wise as Tycho Brahe in lore of the skies. In some way his mother had learned by experience all that Tycho surmised, and, after a long period of hesitation, she was finally persuaded to confide in her son the source of her knowledge. Thus Duracotus learned that his mother was in league with the "daemons of Levania"—the spirits of the moon —whom she could summon upon occasion, and with whom mortals might voyage to the distant land. Upon a certain evening, Duracotus achieved his desire: the time was spring, the moon was crescent and joined with the planet Saturn in the sign of Taurus; the omens were auspicious. "My mother, withdrawing from me into the nearest cross-roads, and uttering a few words loudly . . . returned, and, commanding silence with the palm of her right hand outstretched, sat down near me. Scarcely had we covered our heads with a cloth (as is the custom) when behold, there arose the sound of a voice. . . ."

So ends the first section of the *Somnium*. In spite of the language of legend and superstition, the first part of the tale is clearly based upon Kepler's own life, the allusions so thinly veiled that they can be readily recognized—as Kepler was to learn to his sorrow. The parallel does not, indeed, hold good throughout. The father is a fictional character; Kepler's own father, "ignoble scion of the noble family of Kepler . . . a mercenary of the notorious Duke of Alva," [2] lived only too long after his son's birth. More than once he deserted wife and children, so that the boy grew up in a poverty not far different from that described in his tale. Kepler's mother, however, is truly depicted. Almost illiterate, far below her husband in birth, she was nevertheless a woman of remarkable attainments, a "wise woman" in the true sense of the term. She was also a woman of ungovernable temper, which was finally responsible for the great tragedy of Kepler's life. The early life of Duracotus differed from that of his creator in that Kepler finally managed to secure an education, including training in theology at the University of Tübingen, where he came under the influence of Michael Maestlin, and for the first time came into contact with the revolutionary theories of Copernicus, from which he never departed. Duracotus, as a youth, was apprenticed for five years to Tycho Brahe on the island on which the Danish astronomer established his castle of the heavens, Uraniborg; Kepler himself became Tycho's assistant at a somewhat later age and for a shorter time, when Tycho settled in Prague. But as in the tale, the older Tycho and the younger Kepler worked together for several years, and at Tycho's death in 1601, Kepler fell heir to the rich collection of papers and notes in which Tycho left many of his findings to posterity. Kepler's first monumental work, the *Astronomia Nova,* in which he propounded the first two of his laws, was the result of long work with Tycho on the problem of the planet Mars.

So far the parallelisms, while purposely not exact, are obviously intentional. For my present purposes, only one other biographical fact is necessary. In 1615, largely as a result of her constant quarrels, Kepler's mother was charged with sorcery,

2 *Johann Kepler: a Tercentenary Commemoration of His Life and Work.* Published by the *History of Science Society* (Baltimore, 1931), p. 2.

and came near condemnation.[3] Kepler worked heroically to free her from the charges, and was finally successful after a law suit that lasted five years, during part of which his mother was in prison under ignominious conditions. Shortly after her release, she died.

So much for the similarities between the tale and Kepler's own life. But before indicating what seem to me the important implications, it is necessary to consider the date of composition of the first part of the *Somnium*. The date of publication will offer no assistance, for Kepler left the manuscript unfinished when he died.[4] The only section dated by Kepler himself is the series of notes, which were written between 1620 and 1623. Hence Charles Frisch, the editor of the *Opera Omnia* of Kepler, inclines to that period as the time of composition,[5] though he reminds us of Kepler's interest in the astronomy of the moon, which appears as early as his student days at Tübingen, when he defended certain theses in regard to the moon. The other important German editor, who has made the only modern translation of the *Somnium*, Ludwig Gunther,[6] takes for granted that the work was written in 1609, because of a sentence in Kepler's *Dissertatio cum Nuncio Sidereo,* published in 1610. Here Kepler states that during the previous summer he had applied himself to fundamental problems concerning the moon, though he does not definitely say what he was

[3] The most complete account of the trial is given by Ludwig Gunther, *Ein Hexenprocess: Ein Kapitel aus der Geschichte des dunkelsten Aberglaubens* (Giessen, 1906).

[4] His son-in-law, whom he named as literary executor, endeavored to prepare it for the press, but he too died during his work upon the manuscript. Kepler's son finally acted as editor.

[5] Frisch in his "Proemium Editoris" has collected all references to the *Somnium* in Kepler's other works. He does not actually commit himself as to the date of composition.

[6] *Keplers Traum von Mond* von Ludwig Gunther (Leipzig, 1898), p. x. Gunther follows closely Frisch's evidence, but inclines to date the original composition in 1609 because of the sentence in the *Dissertatio,* which Frisch had quoted. He does not discuss the matter in any detail.

writing on the subject. While it is entirely possible that much of the *Somnium* might have been written in the summer of 1609, it could not then have been written in its present form. Gunther does not see that the moon-world of the *Somnium,* influenced though it was by Plutarch, Lucian, and a host of other early writers, is not the moon-world of the classics, but of Galileo's telescope. There are details which could not possibly have been known to Kepler before the spring of 1610, when the "optick tube" of the Tuscan artist disclosed a new heavens, and Galileo in the *Sidereus Nuncius* announced that the moon was a world, like our world in topography, with irregularities which could not be detected by the naked eye.

It is a remarkable thing that no one of the commentators on the *Somnium* seems to have considered Kepler's own evidence on the subject, presented in one of the notes,[7] nor

---

[7] Note 8, edition 1634, p. 32; Frisch, *Opera Omnia*, pp. 41-2: "Fallor an author [*sic*] Satyrae procacis, cui nomen Conclave Ignatianum, exemplar nactus erat hujus opusculi; pungit enim me nominatim etiam in ipso principio. Nam in progressu miserum Copernicum adducit ad Plutonis tribunal, ad quod, ni fallor, aditus est per Heclae voragines. Vos amici, qui notitiam habetis rerum mearum, & quae mihi causa fuerit peregrinationis proximae in Sueviam, praesertim, si qui vestrum antehac manuscriptum nacti fuerunt, libellum istum, ominosa ista mihi meisq[ue] fuisse censebitis. Nec ego dissentio. Magnum equidem est mortis omen in vulnere lethali inflicto, in veneno epoto; nec minus fuisse videtur cladis domesticae, in propalatione hujus scripti. Credideris scintillam delapsam in materiam aridam; hoc est, exceptas voces istas ab animis intus furvis, furva omnia suspicantibus. Primum quidem exemplar Praga Lipsiam, inde Tubingam perlatum est anno 1611 a Barone a Volckerstorff, ejusque morum & studiorum Magistris. Quantum abest, ut credatis, in Tonstrinis, (praesertim si quibus est ab occupatione Fiolxhildis meae nomen ominosum) in his igitur, confabulatum fuisse de hac mea fabula? Certe equidem ex illa ipsa urbe & domo enati sunt sermones de me ipso calumniosi proxime succede[n]tibus annis: qui excepti ab animis insensis, tandem exarserunt in famam, imperitia & superstitione sufflantibus. Nisi fallor, sic censebitis potuisse & domum meam carere vexatione sexennali, & me peregrinatione annali proxima, nisi somniata praecepta Fiolxhildis hujus violassem. Placuit igitur mihi, somnium hoc meum ulcisci de negocio exhibito, vulgatione libelli: adversariis aliud mercedis erit."

the implications in that note concerning the part played by the *Somnium* in Kepler's own life.[8] To be sure, it is a cryptic note, purposely mysterious in language, intended to be understood only by those of Kepler's contemporaries who had intimate knowledge of his own life. Yet it is clear to anyone who recalls the circumstances under which Kepler's mother was tried for witchcraft. In the note, Kepler is referring to some work of his own which he calls at one time "libellum," at another "manuscriptum." After two sentences which I shall consider later, he says: "You, my friends, who are familiar with my affairs and understand the reason for my late journey into Swabia [9] (especially if any of you had had a look at the manuscript previously), you will understand that that little book, that those happenings, were of evil omen to me and mine. I think so, too. There is indeed a deep foreboding of death in the infliction of a deadly wound, in the drinking of poison; and there seems to have been no less of private tragedy in the circulation of this work. It was really a spark dropped on

[8] There is no indication that any commentator upon the *Somnium* has considered this note, with the single exception of Ludwig Gunther. He translated it in part, *op. cit.*, pp. 27–28, but shows no awareness of its significance. He omits entirely the sentences referring to the *Conclave Ignatii*, evidently considering them either unintelligible or not significant. He translates only a part of the note, and varies the order of Kepler's sentences, so that the final impression is quite different from that Kepler intended. He evidently takes for granted that Kepler was referring to some manuscript which was never published—in spite of Kepler's last sentence, declaring that he intends to publish the manuscript. Neither in the *Traum von Mond* nor in the *Hexenprocess* does Gunther show any awareness of Kepler's suggestion that this work was in part responsible for his mother's trial for witchcraft.

[9] Katherine Kepler, though technically under arrest after 1615, was actually imprisoned only in 1620, when she was taken to the prison in Leonberg. [Gunther, *Ein Hexenprocess*, p. 41.] She was transferred to Gueglingen on August 25 of that year, and kept in the tower in solitary confinement, under unhealthy conditions, in heavy iron chains. [*Ibid.*, p. 47.] Kepler demanded that she be lodged in more humane quarters; from that time on she was kept in the house of the town's gate-keeper, though again chained and guarded by two men. Kepler made a journey to protest against her treatment.

kindling wood—by which I mean those reports, caught by hearts black to the core, filled with dark suspicion." In the next sentence Kepler offers a specific date for the circulation of the manuscript. "The first draft was carried from Prague to Leipzig, and thence to Tubingen in 1611 by Baron Volkerstorff [10] and his tutors." He suggests the result: "Now is it not only too probable that in the barbers' shops—especially those where my name is in bad repute because of the occupation of my Fioxhilda—there was gossip about this story of mine? No doubt at all that from that same city and house lying tittle-tattle came forth about me, me myself, in the years that followed, and that these whispers, harboured by stupid minds, and fanned by ignorance and superstition, blazed out at last into a real story."

Clearly, then, some work of Kepler's, written about 1610, was circulated in manuscript, and, carried into the *"tonstrinae"* —those early predecessors of the coffee-house—fanned a spark already burning, which then blazed up into the fire that almost consumed Kepler and his mother. Reasons for popular distrust of Kepler in 1611 were already legion. As early as 1597 he had become a religious exile. Archduke Ferdinand was an ardent Catholic, who had sworn that he would extinguish heresy. Kepler was an ardent Protestant. At that time he had fled to Hungary, from which he was permitted to return only to be banished again. At this period he came into contact with Tycho Brahe. In addition to religious difficulties, the Kepler family was in constant trouble in every community in which they lived, largely because of the ungovernable temper of "Fiolxhilda," though Johann Kepler seems to have inherited his mother's inability to live in peace with her neighbors. Most of all, Kepler was an avowed "Copernican" in a period in which Copernicanism was suspect. His Copernicanism had been

10 I can find no information about Baron Volkerstorff; Ludwig Gunther is equally at a loss, and mentions [*op. cit.*, p. 27 n.] that this is the only word printed in black-letter in the original edition. Volkerstorff may well have been a former pupil of Kepler's, to whom, in good faith, Kepler had given the manuscript. Kepler was living in Prague at this time.

clearly shown in the *Astronomia Nova*, which, although it is considered one of the great scientific classics today, had no such reputation in its own day. In 1606 he had defended the "absurd" and "impious" theories of Galileo concerning the "new star" of 1604 against the Aristotelians, and championed those theories with such vigor and mastery of logic that to this day the star of 1604 has been called "Kepler's *nova*." In 1610 he published his *Dissertatio* on the same subject. Thus in 1610 Kepler was in disrepute with his neighbors, and was looked upon with suspicion both by Catholics and by Aristotelians. In addition to all these, Kepler in 1610 had written a "manuscriptum" in which he himself had apparently not only described his own mother as a "wise woman," but had declared that she was in frequent communion with the "daemons" of the moon! For the "libellum" which Baron Volkerstorff carried into the barbershops was none other than the *Somnium* itself. In the little fantasy in which he had described his own life, his own experience with Tycho, and particularly the character of his own mother, Kepler unwittingly put into the hands of his enemy the most potent of all charges against his mother, the evidence of her own son. "Caught by hearts black to the core, filled with dark suspicion," this was in truth "a spark dropped upon kindling wood."

If we consider the context, it will be clear that Kepler is speaking of the *Somnium*. It must be remembered that this particular note was written after the death of Kepler's mother, a fact of which Kepler was only too poignantly aware when thirteen years later he annotated his original text. Duracotus is speaking: "My mother was Fiolxhilda, who having lately died, furnished to me freedom for writing, for which I had been yearning. . . . She often said that there were many ruinous haters of the arts who accuse what they fail to understand because of dullness of mind, and hence make laws injurious to the human race; and, condemned by those laws, doubtless not a few have been swallowed up by the pits of Hecla." These words Kepler had evidently written many years before, when his mother was not only alive, but when she was as yet known only as a "wise woman" in the least dangerous sense of that term. Had he heeded her warning and restrained from writing at that particular time, who can tell how different the fortune of the whole Kepler family might have been? Kepler's mother may have been illiterate; but she was truly wise, and from her own limited experience she realized that calumny might befall the person who dared to think otherwise than did his neighbors. In her limited circle she was to suffer persecution; and her son, in his circle was to suffer also for different heretical opinions. Kepler's Fiolxhilda indeed came close to the gates of Hecla. Her son concluded this most personal of all his notes: "So I decided to take revenge for this dream of mine, for the trouble it gave me, by publishing this little book; there will be some other reward for my enemies."

II

I have purposely omitted discussion of the two first sentences in Kepler's note, since they have nothing to do with biographical facts, and lead us from Germany to England. Kepler wrote: "I suspect that the author of that impudent satire, the *Conclave of Ignatius,* had got hold of a copy of this little work, for he pricks me by name in the very beginning. Further on, he brings up poor Copernicus to the judgment seat of Pluto —if I don't mistake, the approach to that is through the yawning chasms of Hecla." On January 24, 1611,[11] the Latin edition of John Donne's *Conclave Ignatii* was entered in the Stationer's Register; on May 18, 1611, an English translation was entered under the title, *Ignatius his Conclave.* Profoundly moved by the implications of the "new astronomy" in the *Anatomy of the World,* written a year later, in the *Conclave Ignatii* Donne was merely pleasantly amused. In this, the most brilliant of his satires, he makes use of the Galilean idea that the moon may be a world, to propose an immediate translation to that world of all the Jesuits who infest this one. The main satire of the *Conclave Ignatii* is concerned with the

[11] Another edition in Latin appeared in 1611 on the continent; *cf.* Geoffrey Keynes, *Bibliography of John Donne* (Cambridge, 1932).

Jesuits. But into an "impudent satire" Donne has introduced passages concerning the "new astronomy," several of which indicate his recent reading of Galileo and of Kepler.

The similarities between Kepler's brief description and John Donne's finished work are close. Within the first few lines of *Ignatius his Conclave* we find a reference to Galileo, "who of late hath summoned the other worlds, the Stars to come nearer to him, and give him an account of themselves." [12] This is closely followed by a reference to "Keppler, who (as himselfe testifies of himselfe) ever since Tycho Brache's death hath received it into his care, that no new thing should be done in heaven without his knowledge." Later there appears in the satire "a certain Mathematitian, which till then had been busied to finde, to deride, to detrude Ptolomey." By his first speech we recognize him: "Are these [doors] shut against me, to whom all the Heavens were ever open, who was a Soule to the Earth, and gave it motion?" With the author we conclude: "By this I knew it was Copernicus."

The similarities are such that it seems practically certain that Kepler was referring to Donne's work. But how could Donne, who published his *Conclave Ignatii* early in 1611, have known of Kepler's manuscript, which, written in 1609 or 1610, was circulating in Germany at this same time? To be sure, if we accept the earlier date of 1609, there is not much difficulty; if, however, Kepler completed his manuscript, as I believe, after Galileo's publication of the *Sidereus Nuncius* in March, 1610—or even after his own telescopic observations in August, 1610—the problem is more difficult. Since no actual evidence seems to be in existence in Donne's published letters, I must for the present limit myself to hypotheses. Certainly John Donne was in touch with various sources through which he rapidly received news of developments in astronomy, for he was well acquainted with the *Sidereus Nuncius* shortly after its publication. So far as Galileo is concerned, the link may

[12] *Ignatius his Conclave* in *Complete Poems and Selected Prose*, edited by John Hayward (Bloomsbury, 1929), pp. 358-9; the later references will be found on pp. 359, 363.

have been Sir Henry Wotton, Donne's close friend and correspondent, who, as we have seen, sent news of the *Sidereus Nuncius* to England. It is conceivable that Kepler's work, circulating in manuscript, may have been sent or brought to Wotton in Italy. While Wotton seems to have met Kepler only much later, in 1620, when he visited the astronomer at Linz, he knew of Kepler's work in this earlier period. However, we may ask whether there was a more direct way in which a copy of the *Somnium* might have reached Donne in England, since if Kepler completed the manuscript only in 1610, the time element is of importance.

Kepler's most exalted acquaintance in England was James I,[13] to whom Kepler sent a copy of his *De Nova Stella* in 1606, with a flattering inscription. That James continued his interest in the astronomer is shown from Kepler's dedication to him in 1619 of the *Harmonice Mundi* and from James's invitation to Kepler in 1620 to come to England as Astronomer Royal. Donne, of course, had many friends at court, from whom he might have heard of the *Somnium* if Kepler sent a copy of the manuscript to His Majesty, as he occasionally sent his published works. Another source, however, seems to me the most likely one. Among Kepler's correspondents in England, as we have seen, was Thomas Hariot. While only a few of the letters they interchanged have been published,[14] those indicate that the two astronomers were in close touch at the period under consideration. In 1608 John Ericksen, who had been for some time with Hariot in England, returned to

13 It seems likely that James's interest in Kepler came about through Tycho Brahe, whom His Majesty had visited at his observatory when James went to Denmark for his bride, in 1589–90. Kepler, in his dedication of the *Harmonice* to James I, refers to the fact that James, while yet a boy, thought the astronomy of Tycho Brahe worthy of the ornaments of his genius.

14 Five letters which passed between Kepler and Hariot have been published in *Joannis Keppleri Aliorumque Epistolae Mutuae* (Francofurti, 1718). Brief digest of the ones to which I refer below are given by Henry Stevens, *Thomas Hariot, the Philosopher, the Mathematician and the Scholar* (London, 1900), pp. 178–180.

Germany, bearing messages to Kepler. In September 1609, Kepler indicates that Ericksen has been again in England with Hariot and is returning to Prague. Ericksen, therefore, was with Kepler in Prague in the autumn of 1609 at the period when Kepler was working seriously upon problems of the moon. What more natural than that, upon his return to Hariot in England, Ericksen should have discussed in detail Kepler's theories about the moon, and that later, when the manuscript was complete, Kepler should have sent a copy either to Ericksen in England—or if Ericksen had again returned to Prague as he seems to have done annually—by Ericksen to Hariot in England? Although no letters between the two are extant for 1610, letters which passed between Hariot and his disciple Lower show how closely Kepler was being studied by the English astronomers. In view of the great interest of both Lower and Hariot in the new theory, what more natural than that Ericksen should have brought back or sent to England a copy of Kepler's manuscript with his latest conclusions on the subject?

It is impossible to tell, from the published correspondence of John Donne, whether he knew Hariot or not. Yet there were many bonds between them. I shall consider for the moment only one of the many men in England closely associated with both who had reason to be interested in such works as the *Somnium*. Henry Percy, ninth Earl of Northumberland, was known everywhere as a man greatly interested in science and semiscience. With Raleigh and Hariot, he was one of the original members of the ill-fated "School of Night," which Shakespeare may have satirized in *Love's Labor's Lost*, in which recent critics have found traces of Shakespeare's criticism of both Raleigh and Hariot.[15] "Deep-searching Northumberland," as Chapman called him—the "Wizard Earl" as he appears to a modern commentator—was a man who seized upon all that was new. Northumberland had been closely associated with Donne at the time of Donne's furtive marriage

[15] See Frances Yates, *A Study of Love's Labour's Lost* (Cambridge, 1936), Chapter VII and *passim*.

to Anne More in 1610. When it became clear that someone must break the news to the father of the bride, Percy was chosen. Early in February, Northumberland proceeded upon his mission, armed with a letter from Donne.

In 1610, Northumberland, with Raleigh, was imprisoned in the Tower. Such commitment, however, did not mean that either was inaccessible. Indeed, during a long part of their detention, both Northumberland and Raleigh continued to carry on their active interest in science, each of them having his laboratory, both keeping in close touch with the outside world. While Raleigh's bonds were tightened for a short time in 1610, there is no evidence that the limitation continued long, nor was a like penalty imposed upon Northumberland. During the whole period of the imprisonment, Hariot was a constant and frequent visitor to both men. That his interest was not confined to Raleigh is to be seen from Hariot's will, in which he bequeathed to Northumberland his most important telescopes, and in which he gave instructions that his papers were to be sent to the Earl.

Here is the sort of connection that could explain how John Donne in England might have seen the manuscript of the *Somnium,* sent or brought to Hariot from Kepler. The *Somnium* was the kind of work in which men like Northumberland and Raleigh and Donne would have been even more interested than in Kepler's technical published works. For the mystery and mysticism in the *Somnium* formed a combination that fascinated members of the "School of Night," and both Raleigh and Northumberland would know that it would have been equally interesting to John Donne, who must have visited his friend in the Tower as Hariot visited Raleigh and Northumberland.

One final point remains: why did Kepler "suspect" that the impudent English satirist had read the manuscript of the *Somnium?* Certainly there is nothing in Donne's straightforward statement in the *Conclave Ignatii* to lead Kepler to that suspicion. Donne merely said: "Kepler, who (as himselfe testifies of himselfe) ever since Tycho Brahe's death hath received it

71

into his care, that no new thing should be done in heaven without his knowledge." But Kepler understood that reference, and knew that Donne was referring not to the *Somnium* but to Kepler's dissertation on the new star—for the main part of the sentence is merely a translation of Kepler's published words.[16] Why, then, the "suspicion"? The answer to that question is important, and not only brings together into a pattern the labyrinthine threads which have been followed, but also explains something which has puzzled critics—the peculiar structure of the *Conclave Ignatii.*

The main body of Donne's work consists of a series of scenes in Hell, with an inquisition upon a number of men famous because of theological disputes. The introduction of the "new astronomy" seems almost irrelevant: references to Galileo and Kepler appear at the very beginning, and to Galileo again close to the end. Neither one of them is a character in the satire, as is Copernicus. Even more curious is the fact that Donne's work begins with a suggestion that he is intending something different from the work published. Donne's first words suggest that he is writing a cosmic voyage, using the device of trance. "I was in an Exstasie," he says,

> My little wandering sportful Soule,
> Ghest, and Companion of my body,

had liberty to wander through all places, and to survey and reckon all the roomes, and all the volumes of the heavens, and to comprehend the situation, the dimensions, the nature, the people, and the policy, both of the swimming Islands, the Planets, and of all those which are fixed in the firmament." Surely this is a synopsis of a cosmic voyage; yet the device is entirely neglected in the work itself. Only in his brief suggestion that the Jesuits are to be transferred to the world in the moon does Donne suggest it again, and even there, there is no real use of the device implied at the beginning, for the moon-world is to be drawn to earth through Galileo's im-

[16] This was first pointed out by Evelyn Simpson, *A Study of the Prose Works of John Donne* (Oxford, 1924), p. 184 n.

proved optic glasses. One sentence alone in this section carries out the idea of a cosmic voyage: "And with the same ease as you passe from the earth to the Moone, you may passe from the Moone to the other starrs, which are also thought to be worlds." The theme of the ecstatic trance, too, completely disappears from the *Conclave,* and is remembered by Donne only in one of the concluding sentences: "And I returned to my body."

The *Conclave Ignatii,* let us remember, was entered in the Stationer's Register in January, 1611. Presumably it was written therefore late in 1610. Let us suppose for the moment that Donne originally wrote it—as in the main it is—merely as a series of dialogues in Hell, a satire on the Jesuits. Its structure is perfectly consistent if we omit the brief introduction and the final reference to the return of the soul to the body. But when the work was ready for the press, let us suppose again that Donne, through Northumberland or Raleigh or Hariot himself, saw the manuscript of the *Somnium,* and realized the rhetorical value of its dream-vision of a cosmic voyage. Time or perhaps the printer would not permit the fundamental changes necessary to recast the whole work into the form of a cosmic voyage. Donne therefore contented himself with the addition of a new introduction and conclusion in which he deliberately adopted the double device of dream and cosmic voyage used by Kepler in the *Somnium,* with the result that the *Conclave of Ignatius* has continued to puzzle critics who have recognized the inconsistency of the two different forms employed by Donne, but who have found no satisfactory explanation for the lack of artistic unity in the finished work.[17]

This, it seems to me, is the explanation of Kepler's "suspicion" that the impudent satirist had "obtained a copy" of his manuscript. Certainly Kepler was in a position to know whether or not it was possible for the English poet to have done so. This explanation seems to me consistent not only

[17] Charles Coffin, *John Donne and the New Philosophy,* discusses at some length, pp. 204 ff., the inconsistency in the structure of the work, and can find no really satisfying explanation, though he suggests various hypotheses.

with Kepler's note and with the structure of the *Conclave Ignatii,* but also with the hypotheses I have attempted to establish as to the date of composition and the circulation of Kepler's manuscript both in Germany and in England. Intent upon his lunar ideas in 1609, Kepler communicated his interest to Ericksen, who in turn brought the word to Hariot and his group in England. But no matter how much of the *Somnium* had been written in 1609, Kepler must have revised it after the publication of the *Sidereus Nuncius,* which offered proof for his theories and gave him certain details which he could not have known in 1609. Sometime in 1610 the manuscript was complete. Circulated on the continent in 1611, it brought tragedy to Kepler and his family. Circulated in England, in 1610, it brought new light to Hariot and to Lower, and helps to explain the close similarity of certain astronomical conclusions reached in England and on the continent. Falling into the hands of the English poet and satirist, the first modern scientific cosmic voyage, written on the continent, caught the imagination of the English poet, who at least suggested, if he did not finally produce, the first modern cosmic voyage in England.

III

The *Somnium,* as I have suggested, is the first modern scientific moon-voyage. Behind it lies a long literary tradition, his debt to which Kepler was quick to realize. In his notes, he discusses at length various works on the subject that had stimulated his imagination, most of all Lucian's *True History* and Plutarch's *De Facie in Orbe Lunae.* Yet with all its debt to the past, the *Somnium* belongs to the modern world. True, Kepler's device for flight to the moon still looks back to the world of the supernatural, not forward to the development of the flying-machine, as do many of the English cosmic voyages written in the seventeenth century. Duracotus learns from the "*Daemon ex Levania,*" who came from the moon at Fiolx-hilda's summons, that mortals may reach the moon only by the assistance of the "daemons." Yet even this section shows

74

the scientific temper of Kepler, for he discusses, both in text and notes, the effect of gravity upon the human body and the "orb of attractive power" of the earth. Like the later writers of modern cosmic voyages, Kepler's imagination plays, too, with the question of the intervening space, the "dark Illimitable ocean, without bound, without dimension" which Milton's Satan faced as he looked out from hell-gates upon the chaos of the new space. As a scientist rather than as a writer of romance, he considers the effect of the rarefied air upon human beings, and his daemons take care to guard against "prohibited respiration . . . by sponges moistened and applied to the nostrils." Once the "attractive orb" of earth is passed, "conveyance becomes easier"; force is no longer necessary. For more than a century this pattern is followed in English cosmic voyages; voyager after voyager enjoys the strange experience of resting upon his wings, or dismounting from his "flying-chariot," to find himself travelling more swiftly than before, without effort, without hunger, thirst, weariness—all ills resulting from the effect of gravity.

But when Kepler's travellers reach the moon, fantasy drops away, and we find ourselves in the new world in the moon, not with a writer of romance such as Godwin, a satirist like Cyrano de Bergerac, nor yet with a poet; our guide is a true scientist. In this respect Kepler's work is almost unique among modern cosmic voyages. Non-scientific writers spent their originality chiefly on ingenious methods of travelling to the moon, and on descriptions of the voyage. Their moon-worlds are, as a rule, conventional utopias, or mere convenient vehicles for satire concerning social and political customs in this world. Kepler, on the other hand, develops in detail the topography of the moon, as Plutarch had presupposed it, as the telescope had shown it. It is a world as strange to us as the new world seemed to Galileo. Seasons, length of day and night, climate, all these are peculiar to Levania. It is divided into two zones, Subvolva and Privola, the first of which enjoys its "Volva" in place of our moon, the second of which is completely devoid of light. In Privolva "night is 15 or 16 of our days long, and dreadful with uninterrupted shadow." On one zone the sun never shines; all things are rigid with cold and frost. In the other the "parching air burns frore." In Subvolva, the climate is somewhat less intolerable, thanks to the Volva. But throughout the whole of Levania, we find extremes to which ours are as nothing, cold more intense, heat more parching. Geographically, the world of the moon is much like our own, save that everything is on an exaggerated scale, the mountains much higher, the fissures and valleys more profound. The life which exists on the moon—and Kepler continues to posit the existence of life in spite of Galileo's denial that water exists on the moon—bears no relation to our life, for Kepler was too good a scientist not to realize the effect of climate and environment upon life. In Subvolva whatever is born is of monstrous size; the life-span of all creatures and plants is brief, since they are often born to die in a single day, springing up to prodigious size while they exist. Here we find no men and women, but creatures who share a "serpentine nature," though some of them are winged, some crawl, some swim in water. Civilization, as we understand it, does not exist; they build no towns, establish no governments. Nomadic creatures, they appear for a short time in the heat of the sun, like lizards basking in tremendous warmth, then disappear either into the seas or into caverns and fissures nature has designed for their protection. A gigantic race of living creatures, they seem to the modern reader reminiscent of a prehistoric world, lunar pterodactyls or ichthyosauri, as, for a moment basking in heat, then creeping into darkness or flying upon prehistoric wing, they disappear forever from the light of Volva, creatures of only a day. The *Somnium* is a dream; but it is a dream with nightmare touches. From this vision of monstrous and grotesque creatures which man is glad he may forget, we gladly wake, to find with the author that the strange book is only part of a dream. Duracotus, Fiolxhilda, daemons and lunar monsters left behind, the author woke "to find my head covered with a cushion, and my body tangled with a rug." The *Somnium* is over.

It is not strange that the moon-world of the *Somnium* should have continued to haunt its readers. It would be difficult to overestimate its effect in England upon the long tradition of cosmic voyages. John Wilkins, in his *Discovery of a New World,* published four years after the *Somnium,* did much to popularize it for English readers. It was known by nearly all the English writers on cosmic voyages—and, indeed, influenced such modern writers as Jules Verne and H. G. Wells. In its own century it was quoted again and again, sometimes seriously, sometimes with amusement. Henry More, in whose philosophical works the influence of Kepler is found more than once, used the moon-world of the *Somnium* as the basis of his minor poem "Insomnium Philosophicum," in which he, too, beheld a vision of another world of light and darkness. Samuel Butler, who satirized the popular interest in a world in the moon, chose the moon-world of the *Somnium* for his description in "The Elephant in the Moon":

> Quoth he—Th' Inhabitants of the Moon,
> Who when the Sun shines hot at Noon,
> Do live in Cellars underground
> Of eight Miles deep and eighty around
> (In which at once they fortify
> Against the Sun and th' Enemy)
> Because their People's civiler
> Than those rude Peasants, that are found
> To live upon the upper Ground,
> Call'd Privolvans, with whom they are
> Perpetually at open War.

Perhaps, in a study which contains so much conjecture and hypothesis, one more may be permitted. Kepler's is the last important moon-voyage to use the old supernatural means of flight to another world; but the last great supernatural flight through space was written by a poet. The flight of Milton's Satan through Chaos draws clearly from the long tradition of cosmic voyages established by Lucian and Plutarch, newly interpreted by Kepler, Godwin, Wilkins, and many others. May there not also be a momentary reminiscence of the *Somnium* in that voyage through space? For the most part in the first two books of *Paradise Lost* Satan is a majestic figure; yet as he forces his way into the chaos of new interplanetary space, he takes on temporarily something of the grotesqueness of Kepler's lunar creatures, as eagerly the Fiend

> O'er bog, or steep, through strait, rough, dense, or rare,
> With head, hands, wings, or feet, pursues his way,
> And sinks, or swims, or wades, or creeps, or flies.

In another, earlier scene in *Paradise Lost,* I believe that there are definite reminiscences of Kepler's moon-world—in Milton's third Hell. The smaller Hell surrounding Pandemonium had its sources, as I have suggested elsewhere.[18] Pandemonium, the second Hell, has also been shown to have had its original. But what of that vaster Hell, stretching out indefinitely, which is explored in Book II by bands of adventurous fallen spirits —more than a continent, for it contains both continents and seas—almost an unknown new world?[19] Certainly only Milton has expressed in poetry the strange grandeur and grotesque picturesqueness of Kepler's world in the moon. Milton's Hell, like Kepler's moon, is a place of "fierce extremes, extremes by change more fierce"; its cold is colder than anything on earth, its heat more torrid; "the parching air burns frore, and heat performs the effect of fire." As Kepler briefly suggests, so Milton describes the frozen world:

> a frozen continent
> Lies dark and wild, beat with perpetual storms
> Of whirlwind and dire hail, which on firm land
> Thaws not, but gathers heap, and ruin seems
> Of ancient pile; all else deep snow and ice.

Kepler's lunar mountains tower to heights more fearful than even the Caucasus; his caverns and fissures are

[18] "Milton's Hell and the Phlegraean Fields," *University of Toronto Quarterly,* VII (1938), pp. 500–513.
[19] *Paradise Lost,* II, 614–628.

> a gulf profound as that Serbonian bog
> Betwixt Damiata and Mount Casius old
> Where armies whole have sunk.

Visitors to Kepler's moon, in short, would have found themselves in just such a world as was discovered by the wandering fallen angels in Milton's Hell:

> Thus roving on
> In confused march forlorn, the adventurous bands
> With shuddering horror pale, and eyes aghast,
> Viewed first their lamentable lot, and found
> No rest. Through many a dark and dreary vale
> They past, and many a region dolorous,
> O'er many a frozen, many a fiery Alp,
> Rocks, caves, lakes, fens, bogs, dens and shades of death—
> A universe of death which God by curse,
> Created evil, for evil only good,
> Where all life dies, death lives, and Nature breeds,
> Perverse, all monstrous, all prodigious things,
> Abominable, unutterable, and worse
> Than fables yet have feigned, or fear conceived,
> Gorgons, and Hydras, and Chimaeras dire.

IV. Milton and the Telescope

WHEN Milton was born, in 1608, Tycho Brahe's "new star" of 1572, the appearance of which may well have startled his grandfather, and Kepler's "new star" of 1604, the excitement over which his father must have remembered, had already become history. Milton was only an infant when, in 1610, Galileo gave the world the first intimation of the greatest astronomical discoveries of the century, and revealed to man the existence of countless new stars, a new conception of the moon and the Milky Way, and the knowledge of four new "planets" of Jupiter. Milton had therefore no such opportunity as John Donne to realize at first hand the excitement caused by these discoveries or to experience the immediate transformation of imagination produced by the first "optic tube." He grew up in a period that gradually came to take the telescope for granted; he lived into an age which became familiar also with the wonders of the microscope, and began to ponder a world of life too minute for the human eye, as Galileo's contemporaries had considered anew the possibility of life in other inhabited worlds beyond sight. Although in youth Milton undoubtedly knew of the telescope, and may even have read the *Sidereus Nuncius,* he was trained under a system of education which paid no attention to contemporary scientific theories and discoveries, and his own tastes and interests were for letters. The astronomical background of his early works was a heritage from the classics, not from science. As a young man, he never knew the excitement of his older contemporaries who, in youth, had read of a new cosmos which almost overnight disrupted the immutable heavens of Aristotle.

Yet every reader of *Paradise Lost* is aware of the fact that Milton's imagination had been stimulated by astronomy, and more than one modern critic has pointed out the extent to which that astronomy was Copernican or Galilean. The problem of his astronomical references has been so frequently discussed that it needs little repetition here, nor am I concerned with what we usually call "the astronomy" of Milton or any other poet—with his acceptance, that is, of the Ptolemaic, the Copernican, the Tychonic, or the Cartesian hypothesis. I am concerned rather with the stimulus of imagination which the telescope produced in the seventeenth century, and the transformation of imagination which resulted from that instrument. In such a study, Milton affords the most remarkable example of the century. Unlike Donne, whose mind also was clearly stirred by implications of the "perspective glass," Milton's imagination, I am persuaded, was stimulated less by books about the new astronomy than by the actual sense experience of celestial observation. As almost in one night Galileo saw a new universe, so Milton on some occasion "viewed all things at one view" through a telescope. Like his own Satan

> Before [his] eyes in sudden view appear
> The secrets of the hoary Deep—a dark
> Illimitable ocean, without bound,
> Without dimension. . . .

That experience he never forgot; it is reflected again and again in his mature work; it stimulated him to reading and to thought; and it made *Paradise Lost* the first modern cosmic poem in which a drama is played against a background of interstellar space.

The early poetry of Milton is the best evidence that before his journey to Italy there had occurred no stimulation of the imagination in astronomical matters such as may be found in Donne in 1611. Although astronomical references are common enough in the *Minor Poems,* there is no significant sentence, no awareness of the ideas of Galileo, Kepler, Bruno. Most of the early figures of speech are merely descriptive: the *sun* appears frequently, but in such lines as these:

> Now while the heaven, by the Sun's team untrod,
> Hath took no print of the approaching light.

The moon shines for him as for any poet of antiquity:

> the wandering moon
> Riding near her highest noon,
> Like one that had been led astray
> Through the heaven's wide pathless way.

The stars that shine upon his youthful poetry are still the stars of Aristotle, undisturbed by the inruption of Tycho's or Kepler's *novae.* They are the "bright morning star, Day's harbinger"; "the star that rose at evening bright"; or the day star that sinks in the ocean bed. Other references are to conventional astrology. His stars are "bending one way their precious influence," such stars as in their malign aspect have influenced that "starred Ethiop Queen." His planets are not the Medicean, but the mediaeval planets which affected men's lives:

> Whose power hath a true consent
> With planet or with element.

The cosmos of the youthful Milton he inherited from the past and apparently did not question. The "starry threshold of Jove's court" is still the boundary of man's world; "bright Spirits" hover "above that high first-moving sphere"; the "celestial Sirens" of Plato "sit upon the nine infolded spheres." There is nothing, in short, in the early poems of Milton to suggest that his mind had been stirred by pondering upon the new astronomy. Indeed, there is one piece of evidence that it had not. The long passage in *Comus,* in which the Lady and Comus, like academic disputants, consider whether Nature is an evidence of superabundance, bidding man pour himself forth with lavish and unrestraining hand, or whether she is a "good Cateress," who teaches frugality, restraint, proportion, anticipates the dialogue of Adam and the Angel in *Paradise Lost* on the same subject. But while argument in the later poem is drawn from various astronomical hypotheses, no such proof occurred to Milton as he pondered the same problem in youth. His illustrations in *Comus* are from Nature as she shows herself in this little world, Comus suggesting that wherever man looks, whether at the vegetation, the sea, or the earth, he sees Nature pouring herself forth, the Lady replying with what is at best a mild form of ethical socialism, concerned only with the difference between "lewdly-pampered Luxury" and the "holy dictate of spare Temperance." In *Paradise Lost,* after astronomical conceptions have entered into Milton's imagination, and Adam finds himself confused between theories which, on the one hand, argue for disproportion and superfluity, on the other, for moderation and restraint, the arguments are drawn entirely from current theories of astronomy. It is seldom that a poet has given us, in the work of his youth and his maturity, two passages which so clearly suggest the difference which years and experience brought in the seventeenth century.

While the poems offer no evidence that Milton had pondered the new astronomy, his early prose indicates that the soil was being prepared for new ideas on such matters.[1] Mil-

---

[1] In Milton's *Sixth Academic Exercise* (*Private Correspondence and Academic Exercises.* translated by Phyllis B. Tillyard, Cambridge, 1932, p. 103) occurs what is evidently a reference to the telescope, in which Milton puns upon the popular title, "perspective-glass":

> And in times long and dark Prospective Glass,
> Fore-saw what future dayes should bring to pass.

So far as I can see, there is no other reference in the early works to the telescope, and none to Galilean astronomy.

ton's college exercises, suggest that he was inclined toward at least a mild academic radicalism. He was among that group at Cambridge who opposed the traditional philosophy. His *Third Academic Exercise* is an attack on the scholastic philosophy and a defense of the sort of studies Bacon had advocated. Since the adherents of the new astronomy were on the whole anti-Aristotelian rather than anti-Ptolemaic,[2] it is significant that Milton shows himself one with the anti-Aristotelians on various other aspects of the quarrel. But it is even more important to notice whom Milton defended than whom he attacked. The *De Idea Platonica* shows him not only cleverly satirizing the literal-minded Aristotelians of the day, but defending the Platonic philosophy. Even more important is his frequently expressed love of the Pythagorean philosophy, for it must be remembered that to many seventeenth-century minds, the discoveries of such men as Copernicus and Galileo were considered important less for novelty than because they brought back the beliefs of Pythagoras; even a cursory reading of Kepler will suggest the extent to which his mysticism was influenced by the supposed "mystick Mathematick" of the Pythagoreans. To Milton in youth Pythagoras seemed "a very god among philosophers" and his *Second Academic Exercise,* "On the Harmony of the Spheres," is filled with a defense of the philosopher against Aristotle, "the rival and constant detractor of Pythagoras and Plato."

There are other passages in the early exercises even more important as showing the direction of Milton's interests. His *Oration in Defense of Learning* contains many sentences

[2] This is a point which, in my opinion, has not been sufficiently stressed by those who have seen in the adherents of the "new astronomy" disciples of Copernicus ranged against disciples of Ptolemy. The student who reads the early work of Kepler, for instance, will observe that his arguments are against Aristotelian rather than Ptolemaic astronomy. The explanation is to be found in the fact that *philosophically* it was Aristotle, not Ptolemy— who was considered primarily as astronomer and mathematician, rather than as philosopher—who had established the conception of the heavens which dominated thought.

showing his interest in the new arts and sciences which were attracting the thoughtful men of his day; the "Ignorance" he attacks is in part the "ignorance of gownsmen," the "sluggish and languid" complacency of the past, which so satisfied men that they felt there was nothing new to learn, the complete dependence upon mediaeval logic and scholastic metaphysics which, declares Milton, is "not an Art at all, but a sinister rock, a quagmire of fallacies, devised to cause shipwreck and pestilence." Here, as in his later *Tractate on Education*, Milton urges the sort of learning which is not barren, which produces, as Bacon would have said, both "Fruit" and "Light." In one passage in these early works, Milton suggests the so-called "Copernican" point of view.[3] There is still another attitude of mind in these academic exercises which, while not specifically concerned with astronomical ideas, was to prove significant in Milton's thinking, and to make his mind receptive to certain implications in the new astronomy. "Let not your mind," he says, "rest content to be bounded and cabined by the limits which encompass the earth, but let it wander beyond the confines of the world." [4] In spite of the checks which he consciously put upon it, Milton's was one of those minds of which he speaks in the *Areopagitica,* "minds that can wander beyond limit and satiety," can play with concepts of time and space, can deal in "those thoughts that wander through eternity." Such minds were peculiarly sensitive to the implications of the new philosophy.[5]

[3] *Seventh Academic Exercise*, p. 108.

[4] *Third Academic Exercise*, p. 72.

[5] Psychologically it is evident that the most important adherents of the "new astronomy," particularly those who, like Campanella and Kepler, attempted to read important philosophical implications into it, possessed this type of imagination. The opposite type of imagination is seen in Bacon, who, as is well known, showed little interest in any of the conceptions of the new astronomy, and who indeed saw in this tendency of human minds which Milton praises, one of the *Idols of the Tribe.* Cf. the passage in the *Novum Organum,* Aphorism 48, beginning "The human understanding is unquiet; it cannot stop or rest, and still presses onward, but in vain. . . .

II

Until the last few years, there has been no question that during his Italian journey Milton visited Galileo, and consequently no reason to doubt that it was Galileo's telescope which disclosed to him the new conception of the heavens and space reflected in *Paradise Lost*. His own statement in the *Areopagitica* that he "found and visited the famous Galileo, grown old a prisoner to the Inquisition," has always been considered sufficient to establish the fact of the visit. In 1918, however, that statement was challenged by S. B. Liljegren [6] as a part of his general attack on Milton's veracity, and his attempt to build up a conception of Milton's character in which the chief characteristics of the poet were egocentricity and an unscrupulous desire for self-aggrandisement. While Liljegren has not succeeded in persuading most critics,[7] his argument cannot be passed over without consideration. Liljegren's most important point is his evidence—based upon documents which he quotes from the great national edition of Galileo, edited by Antonio Favaro—that during the period 1638–39 Galileo was so inaccessible, both because of the sentence of the Inquisition and because of his own health, that approach to him was difficult, almost impossible.[8] Perhaps the best single

[6] S. B. Liljegren, *Studies in Milton*, Lund, 1908.

[7] See the article by Walter Fischer, *Englische Studien* 52. 390–6, with the reply, *ibid.* 54. 358–66; G. Hübener, *Deutsche Literaturzeitung* 40. 150–1; A. Brandl, *Archiv* 138. 246–7; H. Mutschmann, *Beiblatt* 29. 228–35; F. A. Pompen, *Neophilologus* 5. 88–96, with a continuation of the argument, *ibid.* 354–5. Most of these critics are concerned primarily with Liljegren's contention in regard to "Milton and the Pamela Prayer." His argument about Milton and Galileo, a secondary point, has not occasioned much comment.

[8] Liljegren, pp. 25–34. It should be noticed that while Liljegren acknowledges the visit of D. Benedetto Castelli in the autumn of 1638, he lays stress rather upon the difficulties Castelli met than upon the fact that he succeeded in his request; he passes too easily over the visit of Padre Clement in January, 1639; see Favaro, *Le Opere* 18, p. 42. He omits entirely the visits of Vincenzo Viviani and Torricelli in 1639 and 1641; see *Le Opere* 18, pp. 126, 164. In his overemphasis upon the difficulties of Castelli, he neglects to point out sufficiently that the Inquisition may have had reasons

answer that can be made to this argument is that Signor
Favaro himself, who has more intimate knowledge of this
evidence than any other scholar, has found no reason to doubt
Milton's statement.[9] Nor have other Italian critics who have
considered the matter.[10] There is nothing in the other argu-

---

for suspicion of Castelli which did not exist in Milton's case, particularly
if Milton's visit occurred during his first stay in Florence. At that time Milton was completely unknown to the Inquisition; he was merely a young
English traveller, who carried acceptable letters of introduction. Some of
the Italian critics mentioned below agree that Milton might have found
more difficulty in obtaining access to Galileo after his visit to Rome.

[9] Favaro takes the meeting for granted in *Le Opere*, and evidently has
found no reason since to doubt it, since in an article in *Il Giornale d'Italia*,
18 Giugno, 1922, "Galileo e Milton in Arcetri," he surveys some of the recent
important work on Milton, and discusses the visits of Hobbes and Milton to
Galileo.

[10] The chief Italian treatments of Milton in Italy are the following:
Alfredo Reumont in *Archivo storico italiano* 26 (1902). 427 seqq.; Teresa
Guazzaroni, "Giovanni Milton in Italia," Roma, 1902 (Estratto dal *Giornale
Arcadio,* serie 3); Ettore Allodali, *Giovanni Milton e l'Italia*, 1907 (Chap. 2,
"Questione della visita di Milton a Galileo"; cf. also J. G. Robertson, *Mod-
ern Language Review* 2, 1907, 376); Antoni Serao, *Giovanni Milton*, Salerno,
1907 (this work is not biographical, and does not discuss the matter); G. Fer-
rando, "Milton in Toscana," *Illustrazione Italiana*, October, 1925; Anon.,
"Milton a Firenze," *Marzocco*, November 9, 1925; G. N. Giordano-Orsini,
Milton e il suo poema, 1928; D. Angeli, *Giovanni Milton*, 1928. The most
recent Italian work on the subject is *Galileo Galilei* by Giovanni Lattanzi,
which I have not seen, but Lattanzi's position on the subject is clear from a
short article "Gli Ultimi anni di Galileo Galilei" in *Gli Astri,* Giugno-Luglio,
1924, pp. 210–4, for copy of which I am indebted to Signor Abetti of Arcetri.
Signor Abetti, who is in charge of the Galileo collections at Arcetri, has
found no reason to doubt Milton's statement in regard to his visit, as I am
informed by my colleague Miss Emma Detti, who was good enough to discuss the matter with him at my request. The only problem these Italian
critics raise is whether Milton's visit occurred in the autumn or the spring.
Reumont is inclined to believe that Milton would have found more difficulties after his visit to Rome, when his own political and religious views
were known, but considers it certain that the visit took place (cf. p. 19).
Signor Lattanzi in his article, p. 214, quotes a letter which he supposes to
have been written by Milton to Grotius after his visit, in which he speaks
of Galileo "tormentato com' è dalle sue malattie." The letter, however, was
not written by Milton, but by Grotius to Vossius (*Epistola* 964). It is quoted,

ments of Liljegren which deserves or needs consideration—
nothing which does not arise merely from his own conception
of Milton's character. Against his purely hypothetical position,
then, we still have the evidence of Milton's own statement—
evidence which must remain conclusive until better proof is
produced, and we may continue to believe in Milton's visit to
Galileo, as have the poets and artists whose imagination has
reconstructed the event.[11] Whether it was Galileo's telescope

---

·with correct attribution, by Teresa Guazzaroni, in her article, pp. 8–9. In
this connection should be mentioned the series of letters published by
R. Owen, "Milton and Galileo," *Fraser's Magazine* 79 (1869), 678–84, which,
were they genuine, would afford conclusive evidence of Milton's visit. The
letters, supposed to have been written by Milton, Galileo, and Louis XIV,
were from the collection of M. Chasles, and were by him presented to the
Académie des Sciences, and published in *Comptes rendus,* 28 Mars, 1869,
with comments by Elie de Beaumont, *ibid.,* 5 Avril, 1869. Evidently their
authenticity was not doubted at that time; Mr. Owen discusses them seriously, but there seems no reason to believe that they do not belong with
other "Miltonic Myths" discussed by J. Churton Collins, *National Review*
43 (1904).

11 For the benefit of other students who, like myself, may have had
difficulty in tracing the effect of the Milton-Galileo meeting upon Italian
imagination, I may refer to the valuable section on this subject by J. J.
Fahie, *Memorials of Galileo Galilei,* 1929, and add the following information. In 1868 Giacomo Zanella wrote a poem on the subject, "Milton e
Galileo," *Poesie di Giacomo Zanella,* Firenze, 1933, pp. 99–124, in which
he reconstructed imaginatively the meeting. This poem served as inspira-
tion to Annibale Gatti, who *circa* 1877 painted a picture representing the
meeting. The scene of the painting is laid in the Torre del Gallo, instead of
Galileo's house in Arcetri where the meeting probably took place. Various
copies are extant, some showing variations from the original (Fahie, pp.
97–100). For an edition of the original picture, see Giuseppe Palagi, "Milton
e Galileo alla Torre del Gallo, quadretto a olio del Cav. Prof. Annibale
Gatti; descritto e illustrato da Giuseppe Palagi," Firenze, 1877. In 1893
Tito Lessi produced a smaller picture, which, while less ambitious, is more
nearly true to reality, "Milton e Galileo in Arcetri." A reproduction may be
found in the issue of *Gli Astri* referred to above, p. 211; see also the note
of Antonio Favaro in the same issue, p. 217. In 1880 the sculptor Cesare
Aureli produced a marble composition, again following Zanella (Fahie,
pp. 77–80). In his article in *Il Giornale d'Italia,* Professor Favaro describes
this group and devotes a section of his paper to a plea that the statue may

or not is of no consequence, however, to the main contention of this paper. Telescopes were common both in Italy and in England, and Milton must have had many opportunities to survey the heavens at night, before his blindness made vision impossible. Since all his specific references in *Paradise Lost* are to the Florentine, not to the English instrument, one may still insist that, whether Galileo's or another, an Italian "optic glass" first made him conscious of realms of vision and of thought which his youth had never imagined.

Three of Milton's allusions to the telescope in *Paradise Lost* have been so frequently noted that they need little comment here: a specific reference to the "glass of Galileo"; his comparison of Satan's shield to the "optic glass" of the "Tuscan artist" at evening "from the top of Fesole Or in Valdarno"; and his suggestion that the Garden of Eden was

> a spot like which perhaps
> Astronomer in the Sun's lucent orb
> Through his glazed optic tube yet never saw.[12]

There are, in addition, two references to the telescope in *Paradise Regained* less frequently noticed, both of them in scenes in which Satan displays to Christ the kingdoms of the world and the glory thereof. The means Satan employed for that vision did not trouble the writer of the Gospels; but Milton, product of a scientific age, paused to wonder, and concluded:

---

be moved to Arcetri, "la città scientifica fiorentina per sfolgorare al sole di Arcetri dove la storica visita ebbe luogo" as a consecration of friendship between England and Italy. See also "Galileo with Milton at Torre del Gallo," translated by Paul Selver from *The Apostles* of J. S. Machar, *Sewanee Review* 32 (1924). 30-1. I may also mention Solomon Alexander Hart's picture "Milton visiting Galileo in Prison," 1826, and in addition to the English works already well known on the subject, the imaginative picture given by Alfred Noyes in his *Watchers of the Sky,* 1922.

[12] 5. 261-3; 1 287-91; 3. 588-90. Since Milton seems to be referring here to such "sunspots" as those Galileo discovered, this reference also may be said to be associated in his mind with the Italian rather than the English instrument.

By what strange parallax or optic skill
Of vision, multiplied through air, or glass
Of telescope, were curious to inquire.[13]

Satan returns to the same idea when, in the passage which follows, he suggests that Christ may see many things at one view
because "so well I have disposed My aery microscope." [14]

In *Paradise Lost* are to be found the discoveries which,
from the time of the publication of the *Sidereus Nuncius* in
1610 enthralled poetic as well as scientific minds. Here are the
"thousand thousand stars," the sun-spots, and the Milky Way.[15]
The moon appears in *Paradise Lost* as it had in Italy. This is
no longer the moon of conventional poetry—the moon of
Il Penseroso. It is vastly larger—the largest circular body
Milton could think of when he sought an apt comparison
with the shield of Satan.[16] The moon is to Milton as to Galileo
a world much like this earth in its appearance. There are "new

[13] *Paradise Regained* 4. 40–2.

[14] *Ibid.*, 56–7. The use of the word *microscope* here is curious. The term
microscopium or *microscopio* was used in Italy at least as early as 1625.
While microscopes were known in England between 1625 and 1660, they did
not come into common use until after 1660. The first microscopical observations reported to the Royal Society were those of Robert Hooke on March
25, 1663. Clearly, from the *Transactions* of the society, microscopes were
still a novelty at that time. Since Milton was then totally blind, there is no
possibility that he had seen a microscope, and I am inclined to believe that
either he was using the word loosely or that from vague accounts of the
new instrument, he misunderstood its function. In the passage in question,
he seems to be suggesting a combination of a telescope and some supposed
instrument which would show *interiors* as well as exteriors, since Satan
says that by this means Christ may behold "Outside and inside both." This
is an entirely possible interpretation, since the invention of the microscope and telescope precipitated a number of fantastic experiments with
other instruments which were supposed to have strange powers.

[15] *Paradise Lost* 7. 383, 577–81.

[16] I do not mean to say that the conventional moon of poetry does not
appear in *Paradise Lost*. Cf. for example 4. 606–9. The moon seen by Adam
and Eve is the traditional moon of poetry, except in the scene in which
Adam discusses astronomy with the Angel; but the majority of Milton's references are Galilean.

lands, Rivers or mountains in her spotty globe"; [17] "imagined lands and regions in the moon." The Angel ponders the same problem when he questions "if land be there, Fields or inhabitants." There are spots in the moon, the Angel declares:

> Whence in her visage round those spots, unpurged
> Vapours not yet into her substance turned.[18]

Again the Angel suggests the significance of those spots as the seventeenth century understood them:

> Her spots thou seest
> As clouds, and clouds may rain, and rain produce
> Fruit in her softened soil, for some to eat.[19]

Like the disciples of Galileo, also, Milton was impressed with the discovery of the planets of Jupiter, and by the possibility which Kepler had immediately suggested that other planets might also be found to have their unknown attendants:

> and other Suns, perhaps
> With their attendant Moons, thou wilt descry,
> Communicating male and female light,
> Which two great sexes animate the World.[20]

In common with Galileo and many others of the century, too, Milton had been impressed by contemporary theories of

[17] *Paradise Lost* 1. 287–91. In discussing this passage, Allan Gilbert says ("Milton and Galileo," p. 159): "In mentioning 'rivers' Milton is not following Galileo, who held that there was no water on the moon." He bases this statement upon the *Dialogo intorno ai due massimi sistemi del mondo, Le Opere di Galilei,* 1842, 1. 112. But in the *Sidereus Nuncius,* Galileo said (*Sidereal Messenger,* translated by E. S. Carlos, 1880, pp. 19–20): "If any one wishes to revive the old opinion of the Pythagoreans, that the Moon is another Earth, so to say, the brighter portion may very fitly represent the surface of the land, and the darker the expanse of water. Indeed, I have never doubted that if the sphere of the Earth were seen from a distance, when flooded with the Sun's rays, that part of the surface which is land would present itself to view as brighter, and that which is water as darker in comparison."

[18] *Paradise Lost* 5. 419–20.
[19] *Ibid.* 8. 145–7.
[20] *Ibid.* 8. 148–51.

meteors and comets and shooting stars. A nineteenth-century commentator has drawn attention to his observation that meteors are most common in autumn,[21] as Milton suggests in his picture of Uriel's descent:

> Swift as a shooting star
> In autumn thwarts the night.

Comets, too, had interested Milton, perhaps because of the various controversies to which Galileo's theories on comets gave rise, perhaps because he remembered in his childhood the comet of 1618, and had heard from others of the appearance in the year before his birth of "Halley's comet," which startled the early seventeenth century, and was the cause of many pamphlets, ranging from direful prophecy to scientific theory. At least two fine figures of speech in *Paradise Lost* reflect this interest. Satan as he opposes the unknown Death

> like a comet burned,
> That fires the length of Ophiuchus huge
> In the arctic sky, and from his horrid hair
> Shakes pestilence and war.[22]

The last of Milton's majestic figures in the poem is drawn from the same source. To the sad eyes of Adam and Eve

> The brandished sword of God before them blazed,
> Fierce as a comet.[23]

Yet it may be objected that these passages, though they show that Milton had known and pondered the discoveries

21 This observation was made by Professor Mitchell, the astronomer, and is reported in a paper by his sister, Maria Mitchell, published in *Poet-Lore* 6 (1894). 313–323. Professor Mitchell comments, p. 319, "We of this age suppose this was first known since our recollections." Cf. also Milton's figure, 1. 745–6.

22 *Paradise Lost* 2. 708–11. W. T. Lynn, "Comet Referred to by Milton," *Notes and Queries,* series 7, no. 2 (1886), p. 66, suggests that this refers to the appearance of the comet of November, 1664.

23 *Ibid.* 12. 633–4.

of Galileo—as what thoughtful man of his age had not?—
might have been written by anyone who knew of them from
books, that they do not exhibit actual personal experience with
the telescope. There are, however, two characteristics which
make *Paradise Lost* (and in the first instance *Paradise Regained*) unique, characteristics that critics and poets have always felt peculiarly "Miltonic," yet which have never, it seems
to me, been satisfactorily explained. Even a casual reader of
Milton is aware of the vast canvas with which Milton worked,
and on which he displayed his cosmic pictures. I propose to
analyze again some of those familiar passages, seeking to
determine in how far Milton's imagination had been stirred
by the extent of space of the new universe which the telescope
had discovered.

III

One of the peculiarities of Milton's technique is his sense of
perspective. I shall here only raise, because I cannot pretend
to answer, the question: in how far was the new sense of perspective in seventeenth-century art, both pictorial and literary,
the result of the telescope? Certainly during the period in
which the telescope was first impressing the popular mind, we
feel the expansion of space on canvas and in poetry, as in the
century that followed we can detect in descriptive technique
a new feeling for distance. But this is intended for the present
merely as a suggestion. So far as Milton is concerned, the case
is clear. No preceding poet had been able to take us in
imagination to such heights, such vantage points from which,
like Satan or like God, we behold in one glance Heaven, Earth,
Hell, and Space surrounding all. Even when he is not dealing
with cosmic space, Milton in his mature poems loves far views.
Paradise Regained contains a succession of them, all limited
to this world alone, even though the scope of some of them
is such as to stagger comprehension. The "high mountain" to
which Satan led Christ offers at one view a perspective which
includes "a spacious plain," two rivers, with their junction with
the sea, cities; and, adds Milton,

                                 so large
        The prospect was that here and there was room
        For barren desert fountainless and dry.[24]

The topography of the scene is enough to give the needed
impression of extensiveness; but, not content with that, Milton
goes further, suggesting that "turning with easy eye, thou
may'st behold" Assyria, Araxes and the Caspian Lake, Indus,
Euphrates, the Persian Bay, the Arabian Desert, Nineveh,
Babylon, Persepolis, and half a dozen other real and fabulous
places. This is the vastest prospect in *Paradise Regained,* yet
the same general technique is evident, on a lesser scale, in the
vision of Rome, and of Athens. That Milton himself associated
such views with the sense of distance and perspective given by
the telescope is evident from his references in these passages
to the "telescope" and the "aery microscope."

The use of perspective in *Paradise Lost* is at once more
difficult and more subtle. Geography has become cosmography.
But because the scene of *Paradise Lost* is the cosmos, Milton
has all the more reason to use the technique of the telescope
in order to describe the universe which the telescope had
opened to the eyes of his century. Again and again we have
a sensation of the sudden view of far distance, as with Satan
we look "down with wonder at the sudden view Of all this
World at once." Uriel, explaining the scene to the Satan he
does not recognize, sounds to our ears curiously like a seventeenth-century schoolmaster who combines, with a lesson in
theory, practical demonstration through the telescope:

        Look downward on that globe, whose hither side
        With light from hence, though but reflected, shines:
        That place is Earth the seat of Man: that light
        His day, which else, as the other hemisphere,
        Night would invade; but there the neighbouring Moon
        (So call that opposite fair star) her aid
        Timely interposes, and, her monthly round

[24] *Paradise Regained* 3. 262–4. Cf. *Paradise Lost* 11. 377–411 for a similar
prospect from a hill.

> Still ending, still renewing, through mid-heaven,
> With borrowed light her countenance triform
> Hence fills and empties, to enlighten the Earth,
> And in her pale dominion checks the night.[25]

In other scenes of cosmic perspective, however, Milton, for all the strangeness and novelty of the material with which he is dealing, forgets the teacher in the artist. Sometimes it is God himself whom we observe in far-off prospect of the universe:

> Now had the Almighty Father from above,
> From the pure Empyrean where he sits
> High-throned above all highth, bent down his eye,
> His own works and their works at once to view.[26]

More often it is Satan:

>   upon the firm opacous globe
> Of this round World, whose first convex divides
> The luminous inferior Orbs, enclosed
> From Chaos, and the inroad of Darkness old,
> Satan alighted walks. A globe far off
> It seemed; now seems a boundless continent
> Dark, waste, and wild, under the frown of Night
> Starless expos'd.[27]

It is Satan again who, in prospect of Eden, looks sadly from the earth:

> Sometimes towards Eden, which now in his view
> Lay pleasant, his grieved look he fixes sad;
> Sometimes towards Heaven and the full-blazing Sun,
> Which now sat high in his meridian Tower.[28]

Through Satan's eyes we view the most telescopic of all the scenes in *Paradise Lost:* as Satan

> Looks down with wonder at the sudden view
> Of all this World at once . . .
> Round he surveys, (and well might, where he stood
> So high above the circling canopy

25 *Paradise Lost* 3. 722–32.   26 *Ibid.* 3. 56–9.
27 *Ibid.* 3. 418–25.   28 *Ibid.* 4. 27–30.

> Of Night's extended shade) from eastern point
> Of Libra to the fleecy star that bears
> Andromeda far off Atlantic seas
> Beyond the horizon; then from pole to pole
> He views his breadth,—and, without longer pause,
> Down right into the World's first region throws
> His flight precipitant, and winds with ease
> Through the pure marble air his oblique way
> Amongst innumerable stars that shone,
> Stars distant, but nigh-hand seemed other worlds.[29]

It is, too, through Satan's eyes that we first see far off this tiny world of ours, which has become like the other planets which Galileo had discovered—

> This pendent World, in bigness as a star
> Of smallest magnitude close by the moon.[30]

Such a sense of cosmic perspective is as characteristic of Milton as is the so-called "Miltonic style"—for which, indeed, it is in part responsible; yet it is also characteristic of his generation. We do not find it a century before; and though we may find it frequently enough in the century which follows, in the cosmic poems of the eighteenth century familiarity has lost something of the amazement and fascination with which this first generation of men surveyed the new cosmos. Yet even this magnificent sense of perspective was not Milton's greatest heritage from Galileo and his telescope.

IV

"Shakespeare," Professor David Masson used to say in his lectures at Edinburgh, "lived in a world of time, Milton in a universe of space." [31] The distinction Professor Masson felt is the distinction between two worlds—the old and the new; and the profound difference arises from the seventeenth-

---

[29] *Ibid.* 3. 542–66. [30] *Ibid.* 2. 1052–3.

[31] This sentence was quoted to me by President William Allan Neilson, who had been one of Masson's students. Masson only suggests the idea in his *Life of Milton*. See the 1875 edition, 6. 532 and note.

century awareness of the immensity of space. How valid the distinction is will be clear to any student of Shakespeare and Milton, who, considering them merely as reflectors of the thought of their respective periods, observes their obsession with certain dominant conceptions. Of Milton's fascination with *space,* to which *Paradise Lost* bears witness in nearly every book, there is no indication in Shakespeare. And yet that was not because Shakespeare's imagination was not influenced by abstract conceptions. *Time* with Shakespeare is equally an obsession. The use of actual words is perhaps misleading; yet it is at least interesting to observe that the word *space*— according to concordances—occurs in Shakespeare only thirty-two times, always with an obviously limited meaning; *space* to him was little more than "the distance between two objects." The same concordances list more than eight columns of the use of *time.* An *Index to Shakespeare's Thought* [32] makes no reference to his thoughts about space; yet the same index devotes page after page to his thoughts about time, from Rosalind and Orlando's light dialogues on the relativity of time, through familiar references to the "whirligig of time" which brings in its revenges, to the constant reflections on time on the part of more serious characters. Time is to Shakespeare, "the king of men, He's both their parent, and he is their grave." There are the fine lines in the *Rape of Lucrece* beginning: "Time's glory is to calm contending kings," and, as everyone knows, many of Shakespeare's most familiar sonnets deal with the poet's insistent awareness of time. But with the exception of a few dubious lines, there are no passages in Shakespeare that show his mind playing with concepts of space. His world is still bounded by the sphere of the fixed stars, and, indeed, the orb of the moon is the limit of space in his plays. Though travellers' tales could hold Desdemona spellbound, and the geographical world had grown immensely, Puck could "put a girdle round about the earth in forty minutes." Shakespeare's astronomy is still largely astrology; his conception of the order and relation of the heavenly

[32] Cecil Arnold, *An Index to Shakespeare's Thought,* 1880.

bodies, when suggested at all, still conventional mediaevalism. There is no interest in "other worlds." Certainly no vision through a telescope had disturbed his placid cosmos; nor had he heard, as had Milton's generation

> A shout that tore Hell's concave, and beyond
> Frighted the reign of Chaos and old Night.

No reader of *Paradise Lost,* on the other hand, can fail to be aware of the tremendous scale on which it is conceived, or the part which the concept of space plays in its structure.[33] One explanation of the way in which Milton produces this effect is to be found in his conception of the *world,* which, when compared with earlier cosmic poems, indicates the effect of the new astronomy. When Milton uses the term *world* he customarily means not the "little world of man" but the universe. Milton makes much of the difference between this earth as it seems to those who dwell upon it and to those who survey it from afar, to whom it shows its relative unimportance in the cosmic scheme. To Adam and Eve, as to man at all times, earth seems fixed and secure, the center of the universe. At night they survey from their peaceful bower "this fair Moon, And these the gems of Heaven, her starry train." In the morning, they praise in conventional Biblical language "this universal frame, Thus wondrous fair." Satan perceives the difference between the earth as it appears to the angels and to its inhabitants, for, when he finally reaches it, after his first vision from a distance, he finds

> A globe far off
> It seemed; now seems a boundless continent.

But to those who view it from far off—whether God or Satan —and see it in its relation to the vast expanse of space, "this

[33] This was pointed out by Masson, and by Nadal, in his "Cosmogony of *Paradise Lost,*" and has been reiterated by Gilbert in "The Outside Shell of Milton's World." Nevertheless many critics, since Addison, interpret such a line as, "This little world in bigness as a star" as referring merely to the terrestrial globe.

world that seemed Firm land imbosom'd" is but a "punctual spot," a tiny body, merely one of many stars "not unconform to other shining globes." It has shrunk to minute proportions, "a spot, a grain, An atom, with the Firmament compared." It is, "in comparison of Heaven, so small, nor glistering." The Angel, who knows both the world of man and the great cosmos of which it is a tiny part, explains to Adam the vastness of the universe beyond:

> regions to which
> All thy dominion, Adam, is no more
> Than what this garden is to all the earth,
> And all the sea, from one entire globose
> Stretched into longitude.[34]

Beyond the universe of man—even the vastly expanded universe of the telescope which Milton himself had beheld—there stretched in his imagination space. *Space* dominates *Paradise Lost*. We begin to perceive it first through the eyes of Satan as, astounded and momentarily appalled, he gazes into the Chaos that opens beyond the gates of Hell. This is not Satan's first awareness of the extent of the universe. When earlier he warned his followers in Hell of the herculean task which awaited them, he remembered "the dark, unbottomed, infinite Abyss," the "uncouth way," the "vast Abrupt," the "dreadful voyage," as Belial remembered "the wide womb of uncreated Night" in which the fallen angels had so nearly been "swallowed up and lost." There is no exaggeration in Satan's warning of the "void profound Of unessential Night . . . Wide-gaping" which threatens even angelic natures "with utter loss of being . . . plunged in that abortive gulf." He sets out

> with lonely steps to tread
> The unfounded Deep, and through the void immense
> To search, with wandering quest.

The passage in which Milton describes Chaos reflects the *new space* of telescopic astronomy. There was as yet no

[34] *Paradise Lost* 5. 750–4.

vocabulary to express it, and Milton, in common with the astronomers of his day, was driven to a succession of negatives as

> Before their eyes in sudden view appear
> The secrets of the hoary Deep—a dark
> Illimitable ocean, without bound,
> Without dimension; where length, breadth, and highth,
> And time, and place, are lost.[35]

The "wild Abyss" before him, "the womb of Nature and perhaps her grave," is "neither Sea, nor Shore, nor Air, nor Fire." Again and again Milton searches for terms as Satan "with head, hands, wings, or feet, pursues his way." He meets "a vast vacuity"; he springs upward into the "wild expanse"; he forces his way over the "boiling gulf" of the "dark Abyss," until after immense labor he finally approaches the "sacred influence Of Light" where "Nature first begins Her farthest verge." Milton's description of Chaos, both in its vocabulary and its conception, is the first great attempt of English poetry to picture the indefinite the telescope had shown. Many of its details are classical, some are mediaeval, but fundamentally it is a modern Chaos which no mind had conceived before Galileo.

But the description of Chaos is only the beginning of Milton's attempt to depict the new space. We see it through the eyes of God as he "bent down his eye His own works and their works at once to view," and saw in one glance the sanctities of Heaven close about Him, the "Happy Garden" upon earth, "Hell and the gulf between." We realize it again in the further voyages of Satan—voyages which were inherited from and which were to influence the "voyages to the moon" in which the seventeenth century delighted. At one time Satan beholds "Far off the empyreal Heaven"; at another he wanders in the Paradise of Fools in which strong cross winds blow fools "ten

[35] *Ibid.* 2. 890 ff. Cf. also the passages which describe Satan's return, 10. 282–8; 300; 366–71; 397; 470–7.

thousand leagues awry." Finding at last an entrance to earth, Satan upon the lower stair of Heaven's steps "Looks down with wonder at the sudden view Of all this World at once," and "from pole to pole He views his breadth," before he "throws His flight precipitant" downward. Milton's idea of other worlds adds greatly to the expanse of the universe in such passages as these, for we watch Satan at one time winding his "oblique way Amongst innumerable stars" which to men below seemed distant "but nigh hand seemed other worlds." At another time "through the vast ethereal sky" he "sails between worlds and worlds."

It is not alone Satan's voyages which give the reader the sense of space that pervades the whole poem. One need only compare Milton's story of the creation with its original in *Genesis* to realize the expansion of imagination astronomy has produced. The passages to which he has added non-Scriptural details are particularly those that show the creation of the universe rather than those dealing with earth and man. As Christ and his attendant angels survey the Chaos upon which Deity is to impose order, they see it as had Satan at the gates of Hell:

> On Heavenly ground they stood, and from the shore
> They viewed the vast immeasurable Abyss,
> Outrageous as a sea, dark, wasteful, wild,
> Up from the bottom turned by furious winds
> And surging waves.[36]

The first Creation produces the earth: "And Earth, self-balanced, on her centre hung." The firmament that follows the creation of light is diffused

> In circuit to the uttermost convex
> Of this great round.

The Sun and Moon follow, together with the "thousand lesser lights," many of them, even to the phases of Venus and the

36 *Ibid.* 7. 210–4.

Milky Way, in accordance with the new astronomy.[37] The greatness of the descriptive technique in the passage becomes apparent when we realize the subtlety with which Milton suggests the vastness of Space by stressing the *limitation* which Deity imposed "to circumscribe the universe." As the mystic compasses turn "through the vast profundity obscure," the mystic words are spoken:

> 'Thus far extend, thus far thy bounds;
> This be thy just circumference, O World!'[38]

Vast as seems the world, with its light, its firmament, its "thousand, thousand stars," it is yet only a small portion of space. Earlier God had circumscribed for the rebel angels a portion of space that seemed to those still angelic beings to confine them, in spite of the fact that their "adventurous bands" were to discover vast continents of ice and snow, dark and dreary vales, "a gulf profound as that Serbonian bog." The "new-made World," to the angels who beheld its emergence, might seem "of amplitude almost immense," but beyond the world, beyond Hell, even beyond Heaven, in Milton's imagination stretched the "vast unbounded Deep" of Space.

Important as are the scenes of Creation, Milton is still too bound by reverence for the Scriptures to read into them some of the profound ideas which the new concept of space was bringing to men's minds. It remained for the inquiring mind of Adam to raise—if the Angel could not answer—other problems. The long astonomical conversation between Adam and the Angel is concerned with Copernicanism, to be sure, but it also shows the awareness of a vast universe which is post-Copernican. Even to Adam, it is now clear that this Earth is minute in comparison with heaven, an atom when compared with the Firmament:

[37] There is no more charming example of the conjunction of old and new in this age than Milton's introduction into his expansion of *Genesis* of his passage on the phases of Venus, discovered by Galileo (7. 364–9), followed not long afterwards by a Galilean description of the Milky Way (7. 577–81).

[38] *Paradise Lost* 7. 230–1.

> And all her numbered stars, that seem to roll
> Spaces incomprehensible, (for such
> Their distance argues, and their swift return).[39]

He ponders, as had the century, the incredible speed at which these vast bodies must move in incredible space, "incorporeal speed . . . Speed, to describe whose swiftness number fails." The Angel speaks of this also:

> The swiftness of those Circles attribute,
> Though numberless, to his omnipotence,
> That to corporeal substances could add
> Speed almost spiritual.[40]

The vastness of the universe which both the Angel and Adam feel is increased by an idea which the Angel introduces in this particular scene, but which has been recurrent in Milton's mind throughout the poem, as we have already seen—the idea of other inhabited worlds. In this particular passage, it is the Moon which may conceivably be inhabited—"if land be there, Fields and inhabitants." But in other lines in *Paradise Lost*, Milton shows that his mind, as earlier Campanella's, had lingered on the possible existence of other worlds in other stars and planets. Satan considers the possibility as he wends his way

> Amongst innumerable stars, that shone
> Stars distant, but nigh-hand seemed other worlds,
> Or other worlds they seemed, or happy isles,
> Like those Hesperian Gardens famed of old,
> Fortunate fields, and groves, and flowering vales;
> Thrice happy isles! But who dwelt happy there
> He staid not to inquire.[41]

Some such idea is in Satan's mind when, close to Heaven, he pauses to inquire of Uriel, as both of them survey the myriad worlds before them:

> In which of all these shining orbs hath **Man**
> His fixed seat—or fixed seat hath none,

[39] *Ibid*. 8. 19–21. [40] *Ibid*. 8. 107–10. [41] *Ibid*. 3. 565–71.

> But all these shining orbs his choice to dwell . . .
> On whom the great Creator hath bestowed
> Worlds. . . .[42]

Such a universe of habitable worlds is hymned, too, by the chorus of angels on the seventh day when, creation accomplished, they sing not of one world but of many:

> stars
> Numerous, and every star perhaps a world
> Of destined habitation.[43]

True, the Angel, at the end of his astronomical discussion, adds to his suggestions to Adam:

> Dream not of other worlds, what creatures there
> Live, in what state, condition, or degree,[44]

in the same mood in which he assures him that the knowledge of the true astronomical hypothesis is not essential to man. As if in obedience to the command of the Angel, the theme of "other worlds" disappears from *Paradise Lost,* nor does it enter again into any of Milton's works. Yet Milton's mind being what it was—like Adam's, curious in regard to the world about him—we may justly conclude that the apparent coincidence is due less to angelic behest than to the fact that from this time on, he dealt almost exclusively with matters of this world, in the remaining books of *Paradise Lost* and in his last two poems.

Not only are there other existing worlds in Milton's cosmic scheme, but he suggests a still more far-reaching conception which in the age which followed was to develop implications more profound than Milton himself read into it. "Space may produce new worlds," declared Satan to his despondent host upon the lake of Hell. Though Milton did not further develop the suggestion in Satan's speech—for Satan, after all, was little concerned with metaphysics and much with expediency—the idea lies behind several passages in *Paradise Lost.* The "wild Abyss" is "The womb of Nature, and perhaps her grave." The Angel, who, unlike Satan, is concerned with metaphysical

[42] *Ibid.* 3. 668–74.   [43] *Ibid.* 7. 620–2.   [44] *Ibid.* 8. 175–6.

ponderings, goes a step farther, after having suggested to Adam the possibility of life upon the moon. As the telescope of Galileo had discovered satellites around Jupiter, so the Angel suggests there may well be "other Suns with their attendant Moons"

> Communicating male and female light,
> Which two great sexes animate the World,
> Stores in each Orb perhaps with some that live.[45]

Thus having prodigally filled the expanded firmament with suns and stars, having filled the moon with life, and surrounded the suns with attendant moons, the Angel suggests the possibility of future creation, in order that there may not be

> such vast room in Nature unpossessed
> By living soul, desert and desolate.[46]

This is the superabundance and the fertility of Nature which the century was coming to realize, as their conception of life expanded with the expansion of the universe. The development of imagination that has occurred between *Comus* and *Paradise Lost* is obvious. In the youthful poem "Nature" was confined to this earth. Though she might "pour her bounties forth with such a full and unwithdrawing hand," she was still only the productive force which governs the "odours, fruits, and flocks," the "spawn innumerable," the "millions of spinning worms." Her possible "waste fertility" would be shown in an "earth cumbered and the winged air darked with plumes." The last possibility which Comus can conceive is that, unrestrained, she should "bestud with stars" the firmament. The older Milton perceives not only without dismay but even with exultation the vast expansion of a world into a bewildering universe, the possible existence of other inhabited worlds, even the possibility of production of worlds to come.

Yet these are exceptional passages, and no one of them is developed to its full implications. Milton did not in *Paradise*

45 *Ibid.* 8. 148–52.   46 *Ibid.* 8. 153–4.

Lost reach such a conception of the infinity of space as Bruno, nearly a century earlier, nor such an idea of infinite fullness as evidence of Deity as did Leibniz, not much later. Though we may justly say that in comparison with Dante's, Milton's universe has become indefinite, there is here no conception of infinity. Indeed, one of the most remarkable characteristics of Milton's conception of space is his combination of definiteness and indefiniteness. Like his Christ, in the scene of creation, he seems on the one hand enthralled by the "vast immeasurable Abyss," on the other, laboring "to circumscribe This universe." If his "rising World" is "won from the void and formless Infinite," it is nevertheless a measurable world, in which Hell is

> As far removed from God and light of Heaven
> As from the centre thrice to the utmost pole.[47]

True, Milton's angel warns us that in speaking of things infinite, he must speak, as it were, Platonically, and must describe "what surmounts the reach Of human sense" in such terms "as may express them best." Nevertheless, even in his conception of that immeasurable Space which continues beyond the world already created and the worlds to come, Milton does not approach the problem of absolute space as did his Cambridge contemporary Henry More, for example, who at almost the same time was introducing into English thought new concepts of space which were to influence Barrow, Newton, and others. Only at one point in *Paradise Lost* does Milton suggest the problem of Infinite Space and Infinite Deity, the problem which motivates so much of the philosophy of Henry More and of Malebranche. It is tempting to read the words of Milton's Deity

> Boundless the Deep, because I am who fill
> Infinitude; nor vacuous the space. . . .[48]

in the light of contemporary spatial conceptions; and, indeed, considered with More and Malebranche, they may seem to take

[47] *Ibid.* 1. 73-4.   [48] *Ibid.* 7. 168-9.

on new meaning. But so much has already been made of this passage that any Milton student must be aware of the dangers of seeking in any one source the origin of what was probably in Milton's mind a conventional, though complex, theological idea. There is no question that to Milton, God, not Space, was infinite; and no one was more conscious than he of the logical and theological fallacy of making

> Strange contradiction; which to God himself
> Impossible is held, as argument
> Of weakness, not of power.[49]

In the *Treatise of Christian Doctrine*, in which Milton, as theologian, might well have discussed further implications of the *idea of infinity* which were being reflected in contemporary philosophical works, he avoids the whole problem of the nature of space in his discussion both of the creation and of the nature of God. Milton's theology, on the whole, as has been pointed out, draws from a tradition which is antithetic to that which was at least temporarily to triumph in establishing in the seventeenth century a theory of infinite universe as the inevitable expression of infinite Deity, the essence of whose Nature is the overflowing goodness that must show itself in the creation of all possible forms of existence in the created universe. Had he expressed himself on the subject in the *Treatise of Christian Doctrine,* there is little doubt that he would have denied the possibility of infinite space. Yet Milton was first of all a poet; as poet he shows in *Paradise Lost* a momentary imaginative response to certain impressions of the "new astronomy," which, carried to their ultimate conclusion, were inconsistent with his own theological premises. But Milton in *Paradise Lost* was concerned much more deeply with ethics than with metaphysics. Like his Angel, he turns from astronomical implications and from metaphysical considerations of space, to bid Adam, "Think only what concerns thee and thy being." It is enough for him that the expanded universe sug-

[49] *Ibid.* 10. 799–801. Cf. *Treatise of Christian Doctrine,* Chapter 2, section 9.

gests an expansion of Deity; vast though the universe has become, "Heaven's wide circuit" bespeaks for Milton, as for the
Psalmist and the Prophets,

> The Maker's high magnificence, who built
> So spacious, and his line stretched out so far,
> That Man may know he dwells not in his own—
> An edifice too large for him to fill.[50]

Although Milton's mature prose and poetry, then, offers little
to the philosopher seeking new concepts of absolute or infinite
space stimulated by the new astronomy, *Paradise Lost* still
affords a remarkable example of the extent to which telescopic
astronomy effected in an imaginative mind a vast expansion of
the idea of space. Sensitive men of the seventeenth century,
who by actual physical experience of the night sky seen
through an "optic glass" had become aware of "stars that seem
to roll Spaces incomprehensible," did not return to the limited
conception of the universe which they had once taken for
granted. As *Paradise Lost* was affected by the new astronomy,
so in its turn it affected other poets. The impression of space
which Milton achieved is imitated with more or less success by
many poets of the later and the next century. The "sublimity"
of Milton to them was not a matter only of his language, and
his lofty conception of God and Satan, Heaven and Hell, but
even more of his sense of space, the vast reaches of his cosmic
imagination—

> Et sine fine Chaos, et sine fine Deus,
> Et sine fine magis, si quid magis est sine fine. . . .[51]

The eighteenth-century growing "delight in wide views" of
which critics have spoken, has usually been associated with
growing interest in mountains and in mountainous scenery;

---

[50] *Ibid.* 8. 100–4.

[51] Barrow's commendatory verses, prefixed to the second edition. While
I have purposely read into the second line—by omitting the next—an
implication which the poet did not intend, the whole poem indicates the
impression which Barrow had received of Milton's boundless conceptions.

but in the sense of perspective and the awareness of space that enters English writing after *Paradise Lost,* there is in part a direct heritage from Galileo's telescope and in part a heritage from Milton, whose patron goddess was "Urania," and who, even more truly than we have realized, succeeded in portraying in *Paradise Lost* "things unattempted yet in prose or rhyme."

V. The Scientific Background
of Swift's Voyage to Laputa

AMONG the travels of Gulliver, the *Voyage to Laputa* has been most criticized and least understood. There is general agreement that in interest and literary merit it falls short of the first two voyages. It is marked by multiplicity of themes; it is episodic in character. In its reflections upon life and humanity, it lacks the philosophic intuition of the voyages to Lilliput and Brobdingnag and the power of the violent and savage attacks upon mankind in the *Voyage to the Houyhnhnms*. Any reader sensitive to literary values must so far agree with the critics who disparage the tale. But another criticism as constantly brought against the *Voyage to Laputa* cannot be so readily dismissed. Professor W. A. Eddy, one of the chief authorities upon the sources of *Gulliver's Travels*, has implied the usual point of view when he writes:

There seems to be no motive for the story beyond a pointless and not too artfully contrived satire on mathematicians. . . . For this attack on theoretical science I can find no literary source or analogue, and conclude that it must have been inspired by one of Swift's literary ideocyncracies [*sic*]. Attempts have been made to detect allusions to the work of Newton and other contemporary scientists, but these, however successful, cannot greatly increase for us the slight importance of the satire on Laputa.[1]

Three themes in the *Voyage to Laputa* have been particularly censured by modern critics. Some are repelled by the Laputans with their curious combination of mathematics and music and their dread of a comet and the sun.[2] Others are disturbed by the apparent lack of both unity and significance in the Balnibarbians, particularly in the Grand Academy of Lagado.[3] Most of all, the Flying Island has puzzled commentators who dismiss it as a "piece of magical apparatus," a "gratuitous violation of natural laws"[4] which offends the reader's sense of probability.

Yet is it conceivable that Swift, elsewhere so conscious of the unwritten law of probability, should have carelessly violated it in the *Voyage to Laputa* alone? Professor Eddy in a later work has justly said:

The compound of magic and mathematics, of fantasy and logic, of ribaldry and gravity, is a peculiar product of the disciplined yet imaginative mind of Swift. There are two distinct kinds of imagination: one is creative and mystical, the other is constructive and rational. Swift had no command over the faerie architects who decree pleasure domes in Xanadu without regard to the laws of physics. His imagination, like that of Lewis Carroll, had a method in its apparent madness. . . . What seems so lawless is the product of the most rigid law.[5]

Swift's imagination was eclectic; the mark of his genius lay less in original creation than in paradoxical and brilliant combinations of familiar materials. Indeed, one of the sources of his humour to every generation of readers has been the recognition of old and familiar themes treated in novel fashion. Pygmies and giants, animals with the power of speech, have

[1] W. A. Eddy, *Gulliver's Travels: A Critical Study*, Princeton, 1923, p. 158.
[2] *Voyage to Laputa*, in *Gulliver's Travels*, edited by W. A. Eddy, Oxford, 1933, Chap. II. References to the *Voyage to Laputa* are always to this edition.
[3] *Ibid.*, Chap. V.   [4] Eddy, *Critical Study*, p. 158.
[5] Introduction to *Gulliver's Travels*, edited Eddy, p. xviii.

been the perennial stuff of fairy-tale and legend. The novelty in *Gulliver's Travels* lies less in the material than in new combinations and the mood of treatment. The study of the sources of Swift has been particularly rewarding in showing what the "constructive and rational" imagination may do to time-honoured themes. The very fact that the literary and political background of *Gulliver's Travels* has been established so completely leads the inquisitive reader to inquire whether the unrecognized sources of the *Voyage to Laputa* may not be equally capable of verification. If the most assiduous searcher into sources can find "no literary source or analogue" for the peculiar themes in this voyage alone, must not those sources be sought elsewhere than in the literary traditions Swift inherited?

There were other important materials accessible to writers of romance and fantasy in Swift's generation, of which many availed themselves. The attempt of this study will be to show that Swift borrowed for the *Voyage to Laputa* even more than for the other tales, but that the sources of his borrowings were different. The mathematicians who feared the sun and comet, the projectors of the Grand Academy, the Flying Island—these came to Swift almost entirely from contemporary science.[6] The sources for nearly all the theories of the Laputans and the Balnibarbians are to be found in the work of Swift's contemporary scientists and particularly in the *Philosophical Transactions of the Royal Society*.

I

Only during the last few years have students of literature become aware of the part played by the "new science" in the

[6] An occasional commentator has recognized the scientific background of one or another of the details. M. Émile Pons (*Gulliver's Travels* [*Extraits*] . . . *avec une introduction et des notes*, Paris, 1927, p. 204 n.) remarks: "It must be acknowledged, also, that in several of the numerous scientific hints and suggestions contained in these chapters, Swift reveals to us a remarkable and quite unexpected power of divination which is a cause of wonderment for many a scientist of our day."

stimulation of literary imagination in the seventeenth and eighteenth centuries. Among the various themes which have been studied, the relation of men of letters to the Royal Society has proved a rewarding field. Shadwell's *Virtuoso,* for example, takes on entirely new meaning when read in connection with experiments reported by Wilkins, Hooke, Boyle and many others,[7] and the interest of Restoration audiences in such drama becomes clear when we realize the number of men in the audience who had attended meetings of the Royal Society and had heard first-hand reports of the experiments which Shadwell satirizes through "Sir Nicholas Gimcrack." Wilkins, Waller, Dryden, Comenius and Evelyn in England, John Winthrop and Cotton Mather in America, are among those who have been shown to have reflected in their works the discoveries of the Society. Pepys's interest may be traced through frequent references in his diary and letters. Addison and Steele played to a gallery not only of gentlemen but also of ladies interested in science. No reader of *Hudibras* and of the minor poems can fail to be aware of the extent to which Butler, too, found the *virtuosi* an amusing and profitable theme. Foreign visitors to England paid particular attention to "Gresham Colledge," sending home observations, both serious and satiric, upon collections and experiments. It has indeed been suggested that the "early development of Anglomania" during the last years of the seventeenth century and the first quarter of the eighteenth century in France was largely the result of French interest in the Royal Society, particularly in the reports published in the *Philosophical Transactions.*[8]

The widespread interest in scientific discovery among English men of letters was in large part a natural effect of the rapid strides made by science during the seventeenth and eighteenth centuries. More specifically, it was the result of the

[7] See Claude Lloyd, "Shadwell and the Virtuosi," *Publications of the Modern Language Association,* 1929, 44, 472–94.

[8] Minnie M. Miller, "Science and Philosophy as Precursors of the English Influence in France," *Publications of the Modern Language Association,* 1930, 45, 856–96.

attendance at meetings of the Royal Society by men of letters, many of whom claimed the title of *virtuosi,* and of the publication of the *Philosophical Transactions,* which were widely read. In addition to the complete *Transactions,* various abridged editions were published in the early eighteenth century. In 1705 many of the papers were made still more accessible in an edition in three volumes under the title *Miscellanea Curiosa.*[9] Reports of discoveries, inventions and experiments were therefore available in various ways to Jonathan Swift in Ireland, and even more after his return to England, where he "corrected, amended, and augmented" his voyages.[10] The influence

---

[9] *Miscellanea Curiosa. Containing a Collection of Some of the Principal Phenomena in Nature, Accounted for by the Greatest Philosophers of this Age; Being the Most Valuable Discourses, Read and Delivered to the Royal Society, for the Advancement of Physical and Mathematical Knowledge. As also a Collection of Curious Travels, Voyages, Antiquities, and Natural Histories of Countries; Presented to the same Society.* The first volumes appeared at London, 1705–7; an edition, revised and corrected by W. Derham, was published in 1723–7. According to the catalogue of the Library of Congress, the first edition appeared under the auspices of Edmund Halley, the second under the auspices of William Derham.

[10] It is clear, from references in his letters, that Swift was engaged upon *Gulliver's Travels* from at least 1720 until their publication in 1726. He himself points out that he completed, corrected and augmented the travels in 1726, before their publication. Critics disagree about the order of composition, the majority holding that the *Voyage to Laputa* was the earliest and that its imperfections are to be accounted for by the fact that it was written before Swift conceived the work as a whole. One or two critics however have considered the work as the latest of the voyages. The evidence given in this paper tends to bear out the idea that a large part of the *Voyage to Laputa* was a late composition. Several of the experiments which Swift followed most closely were performed as late as 1724. Robert Boyle's complete scientific works appeared in 1725, and though individual papers were available earlier, Swift's close following of Boyle and his many references to him seem to indicate that the complete scientific works were used. The actual dating which Swift himself gives, in the case of the comet —discussed later—and the beginning of the trouble in Balnibarbi, about forty years earlier, both point to 1726 as the most probable date of the composition of the scientific portions of the *Voyage to Laputa.* It would seem that, while Swift was undoubtedly working upon some sections of this voyage in 1724, as references in his letters show, he put aside the more

of the *Philosophical Transactions* upon Swift appears in two ways. These volumes were storehouses of such accounts of travel as those imitated by Swift in *Gulliver's Travels*. In addition they offered him specific sources for his scientific details in the *Voyage to Laputa*.

The various sources of the general idea of such voyages as those of Gulliver have been traced so often and are so obviously part of the great interest in travel that had persisted in England since the time of the earliest voyagers that it seems almost a work of supererogation to suggest the *Philosophical Transactions* as still another source for the main idea of *Gulliver's Travels*. Yet it is at least interesting to see the space devoted in the *Transactions*—particularly between 1700 and 1720 [11]—to accounts of travel. From them Swift may well have gleaned many a suggestion not only for the proper style of Captain Lemuel Gulliver, but also for the pattern of such tales of travel and observation, for such a pattern there was in the actual accounts. The newly discovered islands of the Philippines, reported in the *Transactions*,[12] must have appealed to the creator of Gulliver, who discovered so many islands, some inhabited, some desolate; the Hottentots, as they appear in the accounts sent to the Royal Society, are as curious a people as any discovered by Gulliver.

More specifically, Swift may have picked up from these voyages a hint for his "men who never die," the Struldbrugs. The authors who reported this particular group of travels to the Royal Society showed an almost morbid interest in "antient" men who live too long. Mr. G. Plaxton, a clergyman,

---

technical portions until his return to England; and that the scientific sections of the *Voyage to Laputa* were among the latest of *Gulliver's Travels*.

[11] The abridged edition of the *Transactions* from 1700–1720 devotes a long section (vol. v, Chap. III) to "Travels and Voyages." The third volume of the *Miscellanea Curiosa* is entirely devoted to "Curious Travels, Voyages . . . and Natural Histories."

[12] *Phil. Trans.*, 1708, 26, 189: "An Extract of Two Letters from the Missionary Jesuits, concerning the Discovery of the New Phillippine-Islands, with a Map of the Same."

who seemed to have an uncanny affinity for livings in remote districts, reported from the parsonage of Kinnardsey: "I took the Number of the Inhabitants, and found that every sixth Soul was sixty Years of Age, and upwards; some were 85, and some 90." [13] His next incumbency proved even more remarkable; the number of the aged was much greater and the parishioners lived so long that the Reverend Mr. Plaxton seldom had the pleasure of burying a member of his flock.[14] Cotton Mather sent back from New England "an Account of some Long-Lived Persons there," many of whom were more than a century old.[15] Closer parallels are found in "An Account of the Bramines in the Indies": [16] "It is reported, that upon the Hills by Casmere there are Men that live some hundreds of Years. . . ." Like the Struldbrugs, they have passed beyond curiosity, and beyond interest in life. With Tithonus they seem to have found immortal life but not immortal youth. Gulliver saw the Struldbrugs as "the most mortifying sight I ever beheld. . . . Besides the usual deformities in extreme old age, they acquired an additional ghastliness in proportion to their number of years, which is not to be described." The voyager to the "Bramines" saw his old men in much the same way: "The Penances and Austerities that they undergo are almost incredible; Most of them, through their continual Fastings, and lying upon the parching hot Sand in the Heat of the Sun, are so lean, dry'd and wither'd, that

[13] "Some Natural Observations in Natural History in Shropshire," *Phil. Trans.*, 1707, 25, 2418; Abridged edition, vol. v, ii, pp. 112–15. [In reference to the *Transactions*, we have endeavoured to give references to both the complete and the abridged editions. The first number refers to the complete edition, the second to the abridged, which is more accessible to general readers. We have used the 1749 edition of vols. ii, iii, iv and v, the 1734 edition of vols. vi and vii.]

[14] *Phil. Trans.*, 1707, 25, 2421; vol. v, ii, p. 114.

[15] "An Extract of several Letters from Cotton Mather, D.D.," *Phil. Trans.*, 1714, 29, 62; vol. v, ii, pp. 159–65.

[16] "An Account of the Bramines in the Indies" by J. Marshal, *Phil. Trans.*, 1700–1, 22, 729; vol. v, ii, pp. 165–71.

they look like Skeletons or Shadows, and one can scarce perceive them to breathe, or feel their Pulse beat."

Whether such voyages as those in the *Philosophical Transactions* combined in Swift's mind with literary sources already well established to lead him to the general idea of the travels of Gulliver, we may conjecture, though not prove. That the *Philosophical Transactions*, together with more complete works of the *virtuosi*, were the specific source of Swift's Laputans, his projectors of the Grand Academy of Lagado, and his Flying Island can be proved beyond the possibility of doubt.

II

The section of the *Voyage to Laputa* which deals with the mathematical peculiarities of the Laputans has been generally recognized to be of a piece with others of Swift's pronouncements upon mathematicians.[17] Although several of the critics incline to think that such satire is peculiar to Swift, there is little in the main idea of this section that is unique. Behind the Laputans lay the rapidly growing interest of the seventeenth century in mathematics, embodied in the work of Kepler, Descartes, Leibniz and many others, and a persistent attitude of the seventeenth-century layman toward the "uselessness" of physical and mathematical learning. Bacon's discrimination between "Experiments of Light" and "Experiments of Fruit" had only put into pictorial language a conflict between "pure" and "applied" science. To the layman, and particularly to the satirist of the last quarter of the seventeenth century, when the Royal Society was attracting its greatest attention, the apparent "uselessness" of the new science was a common point of attack. Samuel Butler in *Hudibras* and in minor poems, Shadwell in the *Virtuoso*, Ned Ward in the *London Spy*, William King in the *Dialogues of the Dead*—

17 *Cf.* "Memoirs of Martin Scriblerus" in *Satires and Personal Writings*, by Jonathan Swift, edited by W. A. Eddy, New York, 1933, p. 133: "In his third Voyage he discover'd a whole kingdom of Philosophers, who govern by the Mathematicks."

these and a host of minor writers laughed at the impractical *virtuosi*. A close parallel for Swift's point of view in *Gulliver's Travels* may be found in the *Spectator* papers. In spite of Addison's response to many of the new concepts of the day—his interest in Cartesianism, in Newtonianism, his feeling for the vastly expanded universe of astronomy and biology—he lost no chance for laughter at impractical experimenters and absent-minded mathematicians who in their preoccupation with one subject forgot the world about them.

Swift's Laputans excel in theoretical learning; the abstractions of "higher mathematics" are their meat and drink. They can solve equations—but they cannot build houses, because of the "contempt they bear to practical geometry, which they despise as vulgar and mechanic." Unfortunately their theoretical learning is too abstruse and "too refined for the intellectuals of their workmen." One may wonder whether the passage in which Swift discusses their sharp divergence between theory and practice reflects a point of view suggested by many of the theorists of the day, and expressed by Robert Boyle in these words:

> Let us now consider how far the knowledge of particular qualities, or the physical uses of things, will enable men to perform, philosophically, what is commonly done by manual operation. And here, methinks, 'tis a notable proof of human industry, as well as a great incitement thereto, that philosophy can supply the want to strength, or art, and the head prevent the drudgery of the hand.[18]

If a specific source must be found for Swift's laughter at the uselessness of mathematical learning, it may be discovered in Fontenelle's "Defence" of mathematical and natural philosophy and his insistence that such publications as those of the Royal Society and the French Academy justified themselves. Swift's attitude toward Fontenelle is shown in his earlier *Battle of the Books*, which was largely a reply to Fontenelle's

[18] *The Philosophical Works of the Honourable Robert Boyle, Abridged, methodized and disposed,* by Peter Shaw, London, 1725, vol. i, p. 131. (The pagination in this volume is duplicated in pp. 129–36.)

defence of the "moderns." Another paper by Fontenelle so clearly suggests the position that Swift attacks in the *Voyage to Laputa* that it seems impossible it should not have been in Swift's mind when he wrote. In 1699 Fontenelle, as part of his defence of the "moderns," had upheld mathematical learning in a preface to the *Memoirs of the Royal Academy at Paris*, which was republished as a preface to the *Miscellanea Curiosa* in 1707.[19] The general points attacked by Swift are found in this preface. Fontenelle begins his defence:

> But to what purpose should People become fond of the Mathematical and Natural Philosophy? Of what use are the Transactions of the Academy? These are common Questions, which most do not barely propose as Questions, and it will not be improper to clear them. People very readily call useless, what they do not understand. It is a sort of Revenge; and as the Mathematicks and Natural Philosophy are known but by few, they are generally look'd upon as useless.

He proceeds with a defence of such "useless" knowledge, pointing out on the one hand that supposedly theoretical learning has resulted in practical discoveries, but, on the other hand, defending the intellectual curiosity of mathematicians and natural philosophers as an end in itself. "Altho' the Usefulness of Mathematicks and Natural Philosophy is obscure," he declares, "yet it is real."

The "contempt they bear to practical geometry" is sufficient to explain the miscalculation of the Laputan tailors in making Gulliver's clothes. The mistaking of "a figure in the calculations" may be intended as a satire upon Newton, as has been

---

[19] "A Translation of Part of Monsieur Fontenelle's Preface to the Memoirs of the Royal Academy at Paris, in the Year 1699, treating of the Usefulness of Mathematical Learning," in *Miscellanea Curiosa*, London, 1707, vol. i, Preface. There are striking similarities between the aspects of science defended by Fontenelle and attacked by Swift. Swift possessed in his own library a copy of Fontenelle's *Histoire du Renouvellement de l'Académie Royale*, Amsterdam, 1709. See *A Catalogue of Books: The Library of the late Rev. Dr. Swift*, Dublin, 1745, No. 137, p. 4. This has been republished in *Dean Swift's Library*, by Harold Williams, Cambridge, 1932.

suggested.[20] But like the corresponding passage in the *Voyage to Lilliput,* in which tailors make clothes for the "man mountain," the passage in the *Voyage to Laputa* in which the tailor "first took my altitude by a quadrant" is chiefly a satire upon the current interest in surveying and particularly upon attempts to determine the altitude of the sun, moon, stars and mountains, both lunar and terrestrial, by quadrants and other instruments.[21] Many such papers are included in the *Philosophical Transactions;* the original paper is frequently followed by a rejoinder on the part of another mathematician, pointing out errors in either method or calculation.

But the mathematical interests of the Laputans are not, as a rule, satirized alone; they are included with their interest in music, for in Laputan minds, mathematics and music are one, as they suggest in their clothing, their food and their customs. Here again Swift follows an attitude common enough in the seventeenth century, reflecting Kepler, Descartes, Newton, Leibniz; more specifically, his ideas go back to Dr. John Wallis, who contributed many papers to the Royal Society on the general subject of the analogies between music and mathematics,[22] prefacing his "discoveries" by the suggestions

[20] G. R. Dennis, *Gulliver's Travels,* p. 167, n. 1, has suggested "an error made by Newton's printer in adding a cipher to the distance of the earth from the sun, which drew down some ridicule upon the astronomer."

[21] R. T. Gunther, *Early Science in Oxford,* 1923, vol. i, pp. 345 ff., lists a long series of books and articles on surveying, and discusses, with illustrations, surveying instruments invented in the period, some of them as curious as those found in Laputa. The political significance of such toys is implied in the *Spectator,* No. 262: "The air-pump, the barometer, the quadrant, and the like inventions, were thrown out to those busy spirits [the *virtuosi*], as tubs and barrels are to a whale, that he may let the ship sail on without disturbance, while he diverts himself with those innocent amusements."

[22] See particularly Wallis, "Imperfections in an Organ," *Phil. Trans.,* 1698, 20, 249; vol. i, pp. 612–17. Here Wallis gives an account of the work of his predecessors, and a discussion of "the Degree of Gravity, or Acuteness of the one Sound to that of the other" and of the "Proportions" in music expressed in mathematical formulae. See also "Of the Trembling of Consonant Strings," *Phil. Trans.,* Abridged, vol. i, pp. 606 ff., and "Dr. Wallis's letter

that they "may not be unacceptable to those of the Royal Society, who are Musical and Mathematical." In music and mathematics, many writers of the seventeenth century found the two eternal and immutable verities. Indeed, Christian Huygens went so far as to declare that, no matter how inhabitants of other planets might differ from man in other ways, they must agree in music and geometry, since these are "everywhere immutably the same, and always will be so." [23] The interest of the Laputans in music is not, as has frequently been suggested, a satire upon the interest in Swift's day in opera; the Laputan interest is diametrically opposed and shows the Laputans on the side of those who were resisting the idea that music was a handmaiden to language. Swift's main point is that the Laputans are concerned with the theory, not with the application, of both mathematics and music. Like many of Swift's contemporaries, they expressed their theory of music in mathematical formulae.[24] The Laputans, we are told, express their ideas "perpetually in lines and figures." Such lines and figures—almost equally divided between mathematical and musical symbols—Gulliver saw upon their garments

---

to the Publisher, concerning a new Musical Discovery" in *Phil. Trans.*, 1677, 12, 839.

[23] *C. Hugenii*, ΚΟΣΜΟΘΕΩΡΟΣ, *sive de Terris coelestibus, earumque ornatu, conjecturae*, 1698; editio altera, 1699. The work was translated as *The Celestial Worlds Discover'd: or Conjectures Concerning the Inhabitants, Plants, and Productions of the Worlds in the Planets*, London, 1698. The above reference is to this English translation, p. 86.

[24] For such mathematical interpretations of music, see "The Defects of a Trumpet, and Trumpet Marine" (*Phil. Trans.*, 1692, 17, 559; vol. i, pp. 660 ff.), in which the author writes of the general agreement of "all Writers on the Mathematical Part of Music." Wallis (*ibid.*, 1698, 20, pp. 80, 297; vol. i, pp. 610, 618) discusses the mathematical divisions of the monochord in terms of "Proportions of Gravity" and offers other mathematical analogies. S. Salvetti, in "The Strange Effects reported of Musick in Former Times, examined," points out (vol. i, p. 618) the decadence of contemporary music, in that it was tending away from emphasis on mathematical principles and was coming to be applied merely to "particular Designs of exciting this or that particular Affection, Passion, or Temper of the Mind" —a matter, he feels, to be regretted by true musicians.

and in the King's kitchen, where "all kinds of mathematical and musical instruments" were used to cut the food into "rhomboides" and "cycloids," flutes, fiddles and hautboys. It was entirely natural that, with ideas of beauty founded upon the "Proportions" of music and mathematics, the Laputans should have transferred their figures of speech from one realm to another:

> If they would, for example, praise the beauty of a woman, or any other animal, they describe it by rhombs, circles, parallelograms, ellipses, and other geometrical terms, or by words of art drawn from music.[25]

No specific source is needed for such an idea; and in view of the long succession of predecessors of the Laputans who had found beauty in mathematics and music, no specific source can be posited. Yet the musico-mathematical notions of the Laputans may be conveniently found in a paper by the Reverend T. Salmon on "The Theory of Musick reduced to Arithmetical and Geometrical Progressions." [26] The paper followed an earlier one in which Salmon had reported a "Musical Experiment before the Society," the propositions of which were: "That Music consisted in Proportions, and the more exact the Proportions, the better the Music." In his second paper, Salmon discussed "the Theory of Music, which is but little known in this Age, and the Practice of it which is arriv'd to a very great Excellency," both of which, he suggested, "may be fixed upon the sure Foundations of Mathematical Certainty." He offered in conclusion two tables

[25] *Voyage to Laputa*, p. 191. *Cf.* Fontenelle, "Of the Usefulness of Mathematical Learning," ed. cit.: "A Geometrical Genius is not so confin'd to Geometry but that it may be capable of learning other Sciences. A Tract of Morality, Politicks, or Criticism and even a Piece of Oratory, supposing the Author qualify'd otherwise for those Performances, shall be the better for being composed by a Geometrician. That Order, Perspicuity, Precision and Exactness, which some time since are found in good Books, may originally proceed from that Geometrical Genius, which is now more common than ever."

[26] *Phil. Trans.*, 1705, 24, 2072; vol. iv, i, pp. 469–74.

"wherein Music is set forth, first Arithmetically, and then Geometrically." It required only one more step for the Laputans to apply the certain "Proportions" of music and mathematics to the praise of feminine beauty.

III

More specific satire with more immediate source is found in the sections in which Swift discusses the two predominant prepossessions of the Laputans—their fear of the sun and of a comet.[27] In spite of Swift's suggestion that the Laputans still share astrological fears, he has made them a people whose dread is founded less upon tradition than upon celestial observation. They possess "glasses far excelling ours in goodness," by means of which have extended "their discoveries much further than our astronomers in Europe." [28] They have made important discoveries with their telescopes, none more remarkable than that of the two satellites of Mars—which actually remained hidden from all eyes but those of the Laputans until 1877! [29] They are careful observers, among

[27] *Voyage to Laputa*, pp. 192–3.

[28] *Ibid.*, p. 200. Gulliver adds: "Although their largest telescopes do not exceed three feet, they magnify much more than those of an hundred yards among us, and at the same time show the stars with greater clearness." This passage does not appear in the 1726 edition, but was added in 1727. The addition of the sentence may indicate the current interest in such small instruments. *Cf.* "An Account of a Catadioptrick Telescope, made by Mr. J. Hadley," *Phil. Trans.*, 1723, 32, 303; vol. vi, i, p. 133. The telescope described was six feet long and magnified some 220 times.

[29] M. Pons (*Gulliver's Travels*, ed. cit., p. 234 n.) pays particular attention to this apparently remarkable discovery of Swift's, and points out the similarity not only in the number of satellites but in their periods to the actual discovery made in the nineteenth century. The Laputans found two satellites, "whereof the innermost is distant from the centre of the primary planet exactly three of the diameters, and the outermost five; the former revolves in the space of ten hours, and the latter in twenty one and a half; so that the squares of their periodical times are very near in the same proportion with the cubes of their distance from the centre of Mars, which evidently shows them to be governed by the same law of gravitation, that influences the other heavenly bodies" (*Voyage to Laputa*, pp. 200–1). In spite of a natural desire to agree with M. Pons and Camille Flammarion

whom one would expect to find science rather than superstition. Yet their dread of the sun and of a comet is greater than had been their ancestors', for their fear is more deeply rooted in contemporary science.

Three ideas of the sun particularly disturbed the Laputans:

that the earth, by the continual approaches of the sun towards it, must in course of time be absorbed or swallowed up. That the face of the sun will by degrees be encrusted with its own effluvia, and give no more light to the world. . . . That the sun daily spending its rays without any nutriment to supply them, will at last be wholly consumed and annihilated; which must be attended with the destruction of this earth, and of all the planets that receive their light from it.[30]

Such fears were in no way original with the Laputans. Behind the fear that their planet might fall into the sun lay the potent

---

that this discovery of the satellites of Mars was "second sight" on Swift's part, we are forced to the conclusion that it was only a happy guess. It was inevitable that many writers, scientists and laymen, should have raised the question of the satellites of Mars. (See, for example, Fontenelle, *Plurality of Worlds*, translated Glanvill, London, 1702, pp. 120 ff.) Our own planet was known to have one satellite; Galileo had discovered four about Jupiter; in Swift's time, Cassini had published his conclusions in regard to the five satellites of Saturn. (See *Phil. Trans.*, Abridged, vol. i, pp. 368, 370 and 377; vol. iv, p. 323.) Swift, using no telescope but his imagination, chose two satellites for Mars, the smallest number by which he could easily indicate their obedience to Kepler's laws, a necessity clearly shown him by Cassini; this number fits neatly between the one satellite of the earth and the four of Jupiter. To indicate the Keplerian ratio, he has made one of the simplest of assumptions concerning distances and period, that of 3:5 for the distances, and 10 for the period of the inner satellite. It was not a difficult computation, even for a Swift who was no mathematician, to work out the necessary period of the outer satellite ($3^3 : 5^3 = 10^2 : x$). His trick proved approximately correct—though it might easily have been incorrect. An interesting reply to our theory was made by S. H. Gould, "Gulliver and the Moons of Mars," *Journal of the History of Ideas*, 1945, 6, 91–101.

[30] *Voyage to Laputa*, p. 193. *Cf.* "Memoirs of Scriblerus," ed. cit., p. 136: "A Computation of the Duration of the Sun, and how long it will last before it be burn'd out."

influence of "Britain's justest pride, The boast of human race." Newton's analysis of planetary motion showed that there must exist a nice balance between the velocity with which the earth is falling toward the sun and its tangential velocity at right angles to that fall. Any disturbance of this "due proportion of velocity" would be disastrous. The most obvious possibility of disturbance is the gradual decrease of our tangential velocity, for then the earth's orbit would no longer repeat itself year after year, but would approach the sun with ever-increasing speed, eventually to fall into it. This possibility is recognized in the *Principia* by general calculations of the time required for such falls, and by an estimate of the density of the material in space through which the earth spins and the retarding effect to be expected from it.[31] While Newton's conclusion was, on the whole, an optimistic one that the loss of velocity would be quite inappreciable even for "an immense tract of time," other conclusions were drawn from the same premises. The Laputans might well have found reason for their doubt in Robert Hooke, who, opposing his wave-theory to Newton's theory of light, recognized clearly that there is difficulty in describing the medium which carried these waves and that any imaginable medium would have a retarding effect upon the earth's motion.[32]

Die we must, it would seem, if we are fearful eighteenth-century Laputans. Even if we follow the conclusion of Newton in regard to our earth's falling into the sun, there still remains the warning of the sun-spots and of the consumption of the sun's energy. From the time of Galileo's first observation of sun-spots, astronomers had been concerned to explain these phenomena. During the early years of the eighteenth century

[31] *Philosophiae Naturalis Principia Mathematica*, 1687. Translated into English by Andrew Motte, 1729. References are to the recent revision, *The Mathematical Principles*, edited by Florian Cajori, Stanford, 1934. Cf. particularly Book I, Sections VII and VIII, and Book III, Proposition X, Theorem X.

[32] Cf. "A Discourse of Earthquakes" in *The Posthumous Works of Robert Hooke*, 1705, p. 322.

the *Philosophical Transactions* devoted much attention to the problem of these phenomena. A letter of "Mr. Crabtrie," written in 1640, was revived and republished, and his theory debated, that the spots were "fading Bodies . . . no Stars, but unconstant (in regard of their Generation) and irregular Excrescences arising out of, or proceeding from the Sun's body." [33] At the least, these spots indicated "a Smoak arising out of the Body of the Sun." At the worst, the "Smoak" suggested volcanic action. This point of view was developed in detail by William Derham:

From these preceding Particulars, and their Congruity to what we perceive in our own Globe, I cannot forbear to gather, That the Spots on the Sun are caused by the Eruption of some new Vulcano therein; which at first, pouring out a prodigious Quantity of Smoak, and other opacous Matter, causeth the Spots: And as that fuliginious Matter decayeth and appendeth itself, and the Vulcano at last becomes more torrid and flaming, so the Spots decay and grow to Umbrae, and at last to Faculae; which Faculae I take to be no other than more flaming brighter Parts than any other Parts of the sun. [34]

The Laputans, it would seem, were incorrigible pessimists. Granted we escape falling into the sun, and granted too that the sun-spots indicate only "Smoak," not "Vulcano," our fate will be as surely sealed, if the sun cools or dwindles to a vanishing point. The natural explanation of the heat of the sun, that it is a tremendous burning mass, had been made even more plausible by the discovery of those spots on the sun, which look suspiciously like smoke. Here Hooke was the chief source for concern. "I question not," he wrote, "but that there may be very cogent Arguments drawn to prove, that the Light of this Body of the Sun may be caused by an actual Fire, or Dissolution of the superficial Parts thereof . . . which being proved, or supposed so, all the Appearances that have been

[33] Crabtree's opinion, in a letter to Mr. Gascoigne, is included in the *Phil. Trans.*, 1711, 27, 270; vol. iv, pp. 241 ff. A series of articles on the subject will be found in the *Phil. Trans.*, Abridged, vol. iv, pp. 229–247.

[34] "Spots on the Sun from 1703 to 1708, by Mr. W. Derham," *ibid.*, vol. iv, p. 235.

hitherto taken notice of concerning Clouds, Spots and Blazes, will be very naturally and clearly solved. . . . But some may object and say, that if this were so, certainly the Body of the Sun in so many Thousand Years would have been all consumed, at least it would have grown sensibly less. Suppose this were granted and said, that it has grown some Minutes less since it first began to give Light, none could contradict it by any Observations we have upon record." [35]

Fear of the sun was not all; even greater was the Laputan dread of comets and of one comet in particular. "The earth very narrowly escaped a brush from the tail of the late comet, which would infallibly have reduced it to ashes," Gulliver learned in Laputa. It is not however the "last comet" that terrifies the Laputans so much as one that is to come "which will probably destroy us." Is this mere pointless satire? Swift's imagination here, as so often, is making of the real something apparently unreal. His reference, as every reader of his day must have realized, was not merely to a comet, but to "Halley's comet"—the first comet whose period of return was definitely predicted, with resultant excitement both to literary and to scientific imagination.[36] Thomson, writing only a year later than Swift, shows the same interest when he writes in "Summer":

> Lo! from the dread immensity of space
> Returning with accelerated course,
> The rushing comet to the sun descends:
> And as he sinks below the shading earth,
> With awful train projected o'er the heavens,

[35] "Lectures of Light," in *Posthumous Works of Robert Hooke*, ed. cit., p. 94.

[36] Swift had already satirized the popular excitement occasioned by the prediction of a comet's return in "A True and Faithful Narrative of What Passed in London" (*Satires and Personal Writings*, ed. cit., p. 90): "But on Wednesday morning (I believe to the exact Calculation of Mr. Whiston) the Comet appear'd: For at three Minutes after five, by my own Watch, I saw it. He, indeed, foretold that it would be seen at five Minutes after Five, but as the best Watches may be a Minute or too [sic] slow, I am apt to think his Calculation Just to a Minute."

The guilty nations tremble. . . .
While, from his far excursion through the wilds
Of barren ether, faithful to his time,
They see the blazing wonder rise anew.

In this passage Swift has told us the date of composition of at least part of the *Voyage to Laputa*. The Laputans calculated the return of their comet in "one and thirty years"; thirty-one years after 1726—the date of the first publication of *Gulliver's Travels*—English laymen expected the return of Halley's comet. True, Halley himself had predicted that the comet of 1682 would return not in 1757, as Swift's passage implies, but in 1758; but Halley's prophecy left some reason for doubt. Laymen, then as now, grasped the main point, but neglected the careful mathematics in which Halley corrected a generalization.[37] Seventy-five years had elapsed between the appearance of the comet in 1607 and its reappearance in 1682; years, not days, are important to the layman. The "Mean period" Halley himself calculated at "75 Years and a half." The general public was not at all concerned with the careful table of Halley's "inequalities" nor with his masterful application to his theory of comets of the explanation he had earlier proposed for the deviation from equality in the case of Jupiter and Saturn. As he had suggested that that inequality was the result of the attraction of these planets for each other, in addition to the attraction of the sun for both, so he concluded that the in-

[37] Halley's earliest complete publication on the periods of comets appeared as the "Astronomiae Cometicae Synopsis," published in the *Phil. Trans.* for 1705. An English translation, published at Oxford, appeared in the same year. The latter paper may be found appended to *The Elements of Astronomy, Physical and Geometrical*, by David Gregory, London, 1715, vol. ii, pp. 881–905. The inclusion of this paper in Gregory's volume seems to have escaped the notice of Halley's bibliographers. It is not mentioned in the "Halleiana" in *Correspondence and Papers of Edmond Halley*, edited by E. F. MacPike, Oxford, 1932, pp. 272–8, although the later Latin edition of 1726 in Gregory's volume is noted, p. 278. In the later edition of the *Synopsis*, Halley wrote: "Now it is manifest that two periods of this Comet are finished in 150 Years nearly, and that each alternately, the greater and the less, are compleated in about 76 and 75 Years; wherefore, taking the mean period, to be 75 Years and a half . . ."

equalities in the comet's return might arise from a similar cause. The layman understood only that the comet would appear in approximately seventy-five years; and he vaguely recognized that, if it did, it would put beyond question Newton's theory of gravitation.[38]

In the period of the Renaissance, "Comets importing change of time and states" had brandished their bloody tresses, and predicted "disasters in the sun." But during the seventeenth century, under the impact of the new astronomy, the attitude toward comets began gradually to change, as men questioned whether these strange phenomena too might not prove to have a natural place in the great cosmic scheme. There are indications in almanacs and other popular literature of the day that this was one result of Newton's discoveries. Nevertheless, old beliefs still largely dominated popular imagination. As Swift himself wrote: "Old men and comets have been reverenced for the same reasons; their long beards, and pretenses to foretell events."[39]

The dread of the Laputans rested however less upon such superstition than upon scientific discovery. With their telescopes they had "observed ninety-three different comets, and settled their periods with great exactness."[40] If therefore they

[38] Halley himself was incorrect in his computations, as events proved, since the theory of perturbations was not sufficiently advanced for him to make exact prophecy; in addition, Uranus and Neptune were unknown to his generation. The comet passed perihelion on March 13, 1759, though it was observed on Christmas night, 1758.

[39] Prose Works, edited Temple Scott, London, 1922, vol. i, p. 281.

[40] Voyage to Laputa, p. 201. This is characteristic Laputan exaggeration. No such number of comets had been observed; the periods of only three had been fully calculated by Halley. Gulliver adds characteristically: "It is much to be wished that their observations were made public, whereby the theory of comets, which at present is very lame and defective, might be brought to the same perfection with other parts of astronomy." Among the advantages Gulliver at first thought might come from the immortality of the Struldbrugs, he suggested (p. 250): "What wonderful discoveries should we make in astronomy, by outliving and confirming our own predictions, by observing the progress and return of comets, with the changes of motion in the sun, moon, and stars." One of the "Philosophi-

feared that a comet "one and thirty years hence" would destroy them, they must have had scientific grounds for their belief. The basis for their fear was implied even in Halley's earlier *Synopsis,* in connection with his discussion of the approach of various comets to the earth. His paper concluded with the statement: "But what might be the Consequences of so near an appulse; or of a contact, or, lastly, of a shock of the Celestial Bodies, (which is by no means impossible to come to pass,) I leave to be discussed by the Studious of Physical matters." In his later amplification of the *Synopsis,*[41] Halley went further and expanded this section in connection with the comet of 1680:

Now this Comet, in that part of its Orbit in which it descended towards the Sun, came so near the paths of all the Planets, that if by chance it had happened to meet any one of the Planets passing by, it must have produced very sensible effects, and the motion of the Comet would have suffered the greatest disturbance. In such case the plane and species of its Ellipsis and its periodic Time would have been very much changed, especially from meeting with Jupiter. In the late descent, the true path of this Comet left the Orbits of Saturn and Jupiter below itself a little towards the South: It approached much nearer to the paths of Venus and Mercury, and much nearer still to that of Mars. But at it was passing thro' the plane of the Ecliptic, viz., to the southern Node, it came so near the path of the Earth, that had it come towards the Sun thirty one days later than it did, it had scarce left our Globe one semidiameter of the Sun towards the North: And without doubt by its centripetal force (which with the great Newton I suppose proportional to the bulk or quantity of matter in the

cal and Mathematical Works" of Martin Scriblerus was "Tide-Tables for a Comet, that is to approximate towards the Earth."

[41] There is no mention in the "Halleiana" referred to above of the fact that Halley published a later edition of the *Synopsis,* with several corrections and additions. Such a paper appeared however in *Astronomical Tables with Praecepts Both in English and Latin, For Computing the Places of the Sun, Moon, Planets, and Comets,* by Edmond Halley, London, 1752. The volume is not paginated. The editor states that the tables were "sent to the press in 1717 and printed off in 1719." The chief changes are in the tables; the rest follows the original until the conclusion, to which we refer, which is expanded.

Comet), it would have produced some change in the situation and species of the Earth's Orbit, and in the length of the year. But may the good GOD avert such a shock or contact of such great Bodies moving with such forces (which however is manifestly by no means impossible), lest this most beautiful order of things be intirely destroyed and reduced into its antient chaos.

Although this suggestion alone would have been sufficient to explain the Laputans' dread of the comet, there is little doubt that popular imagination was even more deeply stirred by another paper which Halley presented to the Royal Society —on the subject of Noah and the Flood. This was one of many papers published in the period by important men of science in which an attempt was made to explain difficult passages in Scripture in such a way as to keep the reverence for the Bible, yet make it consistent with modern scientific thought. Straining at the gnat, Halley and others swallowed the Deluge.[42] In an earlier version of the paper, read before the Royal Society in 1694,[43] Halley had suggested "the casual Choc of a Comet, or other transient Body" as "an Expedient to change instantly the Poles and Diurnal Rotation of the Globe." But in the later paper he went further: "At that Time," he says, "I did not consider the great Agitation such a Choc must necessarily occasion in the Sea." Halley's description of the probable consequences of such a "Choc" was sufficient to strike terror into braver hearts than those of the Laputans. He visualizes the Deluge

[42] Cf. "Memoirs of Scriblerus," ed. cit., p. 135: "To him we owe all the observations of the Parallax of the Pole-Star, and all the new Theories of the Deluge."

[43] "Some Considerations about the Cause of the Universal Deluge," Phil. Trans., Abridged, vol. vi, ii, pp. 1–5. The original paper, which differed in several important respects, was read before the Society on December 12, 1694. The later paper was read in 1724. To some extent the ideas suggested in the earlier paper were intended as a confutation of Thomas Burnet's Sacred Theory of the Earth, Londini, 1681-2, 2 vols., a copy of which, containing annotations in his own hand, was in Swift's library. See A Catalogue of Books, ed. cit., No. 375.

raising up Mountains where none were before, mixing the Elements into such a Heap as the Poets describe the Old Chaos; for such a Choc impelling the solid Parts would occasion the Waters, and all fluid Substances that were unconfined, as the Sea is, with one Impetus to run violently towards that Part of the Globe were [sic] the Blow was received; and that with Force sufficient to rake with it the whole Bottom of the Ocean, and to carry it upon the Land; heaping up into Mountains those earthy Parts it had born away with it, in those Places where the opposite Waves balance each other, *miscens ima summis*.

Thus Halley, discovering that the comets, like the stars in their courses, obey the universal law of gravitation, established in 1705 the point of view that was to free men from their long dread of "those stars with trains of fire and dews of blood"; but through a few sentences in a paper in which he announced the law of comets, and, most of all, through republishing in 1724 a paper largely written thirty years before on the subject of that Deluge weathered only by an ark, put into the minds of the Laputans—and many of Swift's contemporaries—a greater dread, of the complete annihilation of this globe which we inhabit.[44] Small wonder that in the morning the Laputans exchanged no trivial greetings. "The first question is about the sun's health, how he looked at his setting and rising, and what hopes they have to avoid the stroke of the approaching comet." Like children who have listened to tales of hobgoblins, the Laputans "dared not go to bed for fear."

[44] The Laputans may also have found reason for their fear of a collision in Newton's passage in the *Principia*, Book III, Proposition XLII, Problem XXII: "The comet which appeared in the year 1680 was in its perihelion less distant from the sun than by a sixth part of the sun's diameter; and because of its extreme velocity in that proximity to the sun, and some density of the sun's atmosphere, it must have suffered some resistance and retardation; and therefore, being attracted something nearer to the sun in every revolution, will at last fall down upon the body of the sun." After a discussion of the appearance and disappearance of *novae*, Newton concludes: "The vapours which arise from the sun, the fixed stars, and the tails of the comets, may meet at last with, and fall into, the atmospheres of the planets by their gravity. . . ."

IV

After his visit to the Laputans, Gulliver descended to the mainland, Balnibarbi, and proceeded at once to the capital Lagado, whose Grand Academy was to prove one of the chief interests of his voyage. He was impressed, both in town and country, by many "odd appearances." The houses were "very strangely built and most of them out of repair." Though the fields were filled with labourers and the soil appeared excellent, he saw neither corn nor grass. As he journeyed with the "great lord Munodi" to his country estate, he observed "the several methods used by farmers in managing their lands, which to me were wholly unaccountable." Only in the privacy of the country-house did Munodi explain to Gulliver the source of the evident difference between his own well-ordered estate and the "ill contrived" buildings and "unhappily cultivated" fields of the rest of Balnibarbi. In so far as this section of the *Voyage to Laputa* has been studied, critics have taken for granted that the source of its satire was in contemporary politics. Balnibarbi is England, or more often Ireland, with its houses out of repair, its fields badly cultivated, its people in misery and want. Yet, though Swift undoubtedly intended some such meaning, there is another sort of satire here also, which leads backward to Swift's part in that particular chapter of the long warfare of "ancient" and "modern" which Swift himself called *The Battle of the Books*.

The aspect of the old quarrel reflected in the *Voyage to Laputa* is not the controversy between "ancient" and "modern" literature, but the broader implications of the battle which in England had become largely a scientific controversy. Munodi, with whom alone the conservative Gulliver found sympathy, is clearly an "ancient" and for that reason ridiculed by his "modern" neighbours. His sympathy with the party of the "ancients" is shown most immediately in his surroundings. His house is "a noble structure, built according to the best rules of ancient architecture" and therefore, Swift slyly suggests, still standing! "The fountains, gardens, walks, avenues, and groves were all disposed with exact judgment and taste." "Everything about him was magnificent, regular, and polite." "He was content to go on in the old forms, to live in the houses his ancestors had built, and act as they had in every part of life without innovation." Yet in the eyes of his countrymen he was not only a failure, but also an enemy to progress, an "ill commonwealth's man." His example would have been considered dangerous, had it not been that that example was followed only by "such as were old, and wilful, and weak like himself." So strong was the pressure of public opinion that Munodi sadly faced the necessity of tearing down his noble structure to rebuild in the present ill mode.

The trouble had begun, he tells Gulliver, "about forty years before," when certain Balnibarbians had visited progressive Laputa and, falling under the spell of Laputan philosophy, "came back with a very small smattering in mathematics, but full of volatile spirits acquired in that airy region." Here, as so often, Swift's figures are significant. "About forty years before" the composition of the *Voyage to Laputa*—thirty-nine years, if we accept 1726 as the year of composition of these sections—the first gun in the *Battle of the Books* had been fired by Charles Perrault's *Siècle de Louis le Grand,* followed the next year by Fontenelle's *Digression sur les Anciens et les Modernes* and the first volume of Perrault's *Parallèle des Anciens et des Modernes,* as a result of which Sir William Temple wrote his essay "On Ancient and Modern Learning" and ultimately drew Swift into the combat. From the time of their visit to Laputa, the Balnibarbians "fell into schemes of putting all arts, sciences, languages, and mechanics upon a new foot." Most of all, they had established an "Academy of Projectors" which had come to dominate the nation, as the Royal Society dominated England. Its "professors contrive new rules and methods of agriculture and building, and new instruments and tools for all trades and manufactures." [45] They

[45] *Voyage to Laputa*, p. 200. The terminology here recalls that of the Royal Society. Cf. *The History of the Royal Society of London, For the*

promised a new Utopia if their methods were followed. True, their magnificent projects had not been brought to perfection at the time of Gulliver's visit; but with Baconian optimism, they persisted in their prosecution of schemes to reform the kingdom by science, centred in a later "Salomon's House."

V

It has been generally recognized that in the Grand Academy of Lagado, Swift was following a fashion, common enough in literature of the seventeenth and eighteenth centuries, of satirizing academicians in general and the Royal Society in particular; but the full extent of that satire has not been appreciated. Long before the incorporation of the Royal Society, Rabelais had introduced a somewhat similar passage in his Court of Queen Whim; Joseph Hall in 1610 described another such Academy in his *Mundus Alter et Idem*. Bacon's enthusiastic proposals for his scientific Academy in the *New Atlantis* offered new fuel. The establishment of the Royal Society tended to make specific former general satire. From the time of Samuel Sorbière's visit to London in 1663,[46] journeys, whether "real" or "philosophical," tended to include accounts of academies which are usually only thinly veiled pictures of the Royal Society. Ned Ward's tour of London, described in the *London Spy* in 1698,[47] led him to Gresham College as well as to the "Colledge of Physicians"; in both institutions he

---

Improving of Natural Knowledge, by Tho. Sprat, 3rd edn., London, 1722, p. 190: "They have propounded the composing a Catalogue of all Trades, Works, and Manufactures, wherein Men are employ'd . . . by taking notice of all the physical Receipts or Secrets, the Instruments, Tools, and Engines . . . and whatever else belongs to the Operations of all Trades." The interest of the Society in "instruments and tools . . . trades and manufactures" is clear throughout.

[46] *Cf.* Vincent Guilloton, "Autour de la Relation de Samuel Sorbière en Angleterre," in *Smith College Studies in Modern Languages,* 1930, 11, No. 4, 1–29.

[47] *The London Spy. Compleat in Eighteen Parts,* by Ned Ward, with an Introduction by Ralph Straus, London, 1924, pp. 50 ff. and 125 ff.

examined the supposed "rarities" and "Philosophical Toys."
From both he went away with a poor opinion of scientists, who
seemed to him only less mad than the inmates of the lunatic
asylum which he also visited. Dr. Martin Lister in the account
of his journey to Paris [48] paid tribute to both French and
English scientists. The subsequent parody by William King in
his *Journey to London* [49] in the same year introduces satirically
the theme of visits to the *virtuosi*. King carries his attack further in the Ninth Dialogue of the *Dialogues of the Dead* and
in the *Transactioneer*.[50] Even closer similarities to Swift's
Academy may be found in Tom Brown's "Philosophical or
Virtuosi Country" in *Amusements Serious and Comical,* and in
Ludwig Holberg's *Journey to a World Underground*. Swift
hardly exaggerated when he said through the lips of Munodi:
"There is not a town of any consequence . . . without such an
academy." While there is no doubt that some of the details
in the *Voyage to Laputa* reflect such earlier works as that of
Rabelais, nevertheless Swift's Grand Academy of Lagado was
drawn rather from life than from literature.

On December 13, 1710, Swift himself had paid a visit to
Gresham College. With many other visitors, his itinerary included other institutions often grouped together in the memory
of travellers to London: "then to Bedlam; then dined at the
Chophouse behind the Exchange; then to Gresham College
(but the keeper was not at home), and concluded the night at
the puppet-show. . . ." [51] Puppet-shows, lunatic asylums, col-

48 *A Journey to Paris in the Year 1698,* London, 1698, pp. 78 ff. and
passim.

49 *A Journey to London. In the Year 1698. After the Ingenious Method
of that made by Dr. Martin L. . . . to Paris, in the same year* in *Miscellanies in Prose and Verse,* by William King, London, 1705, pp. 224 ff.
and *passim.* A modern edition may be found in *A Miscellany of the Wits:
Select Pieces by William King, D.C.L., John Arbuthnot, M.D., and other
Hands,* with an Introduction by K. N. Colville, London, 1920, pp. 15 ff.
and *passim.*

50 *Miscellanies in Prose and Verse,* ed. cit., pp. 324–38; *A Miscellany of
the Wits,* ed. cit., pp. 69–80; *The Transactioneer, with some of his Philosophical Fancies, in Two Dialogues,* by William King, London, 1700.

51 *Journal to Stella* in *Prose Works,* edited Temple Scott, vol. ii, p. 72.

leges for the advancement of research—they were all one to the satirists of that generation. If, in spite of the absence of the keeper, Swift saw any of the collections of the Royal Society, we may perhaps detect reminiscences of his visit in his later references to the loadstone of the Flying Island and his brief suggestion of projectors who were "petrifying the hoofs of a living horse to preserve them from foundering." The collection of petrified objects belonging to the Royal Society was shown with pride to visitors and was known throughout Europe. One section of their earliest catalogue was devoted to "Animal Bodies Petrified," another to "Vegetable Bodies Petrified." Yet on the whole Swift's Academy reflects less Swift's own visit than accounts in the *Philosophical Transactions*.

In his account of the Grand Academy, Swift first describes the institution briefly: "This Academy is not an entire single building, but a continuation of several houses on both sides of a street, which growing waste was purchased and applied to that use." Gulliver later estimates that there were at least five hundred rooms in the institution. This is certainly not the Royal Society as it appeared in the years when Swift was at work on *Gulliver's Travels*. Yet it is possible that this is a sly dig at the Society's ambition for greatly expanded quarters, which threatened to divide the Council into two factions.[52]

52 From the time that the President, Newton, had declared it necessary that they "have a being of their own" (C. R. Weld, *History of the Royal Society*, London, 1848, vol. i, p. 387), one of the most persistent problems reflected in the records is that of moving to larger quarters. When in 1705 the Council received word from the Mercers' Company that the latter had decided "not to grant the Society any room at all" (*ibid.*), purchase of property became imperative. In 1710 two houses in Crane Court in Fleet Street were bought, though the Council was far from unanimous in its decision as a pamphlet of the day indicates (*An Account of the late Proceedings in the Council of the Royal Society, in order to remove from Gresham College into Crane Court in Fleet Street*, London, 1710). During the next few years the Society came to pride itself upon its increasing importance. On December 15, 1710, the Society was appointed Visitors and Directors of the Royal Observatory at Greenwich (Weld, *op. cit.*, vol. i, pp. 400 ff.). Its property, received by deed and

Some members continued to entertain the noble ambition, proposed by Bacon in his description of "Salomon's House," not only of "great and spacious houses," but of "deep caves . . . high towers," great lakes and artificial wells, orchards, gardens, parks and enclosures.

The members of the Academy whom Gulliver encountered were of various groups—experimental scientists, "projectors in speculative learning," professors in the "school of languages" and politicians. Since our concern is with the scientific background of the voyage and particularly its relation to the *Philosophical Transactions,* we may limit ourselves to the first group of experimentalists. The experiments of Swift's projectors have impressed literary historians chiefly by their apparent exaggeration and have been dismissed as so obviously impossible that they seem grotesque rather than humorous. Swift, the critics say, "simply tortured his memory and his fancy to invent or recall grotesque illustrations of scientific pedantry." [53] Yet there was humour in these passages when they were written, and humour of a sort particularly popular today. Swift's is the *reductio ad absurdum* frequently employed by modern satirists who reduce to nonsense scientific papers and doctoral dissertations, not by inventing unreal subjects of research, but—more devastatingly—by quoting actual titles of papers and theses. What, asks a modern reader, could be more absurd than "A Study of the Bacteria Found in a Dirty Shirt"? Removed from its context, read by laymen instead of scientists, the real serves often as a more powerful weapon against scientific research than anything invented by fancy. Such is Swift's technique. For the most part he simply set down before his readers experiments actually performed by members of the Royal Society, more preposterous to the layman than anything imagination could invent and more devastating in their

bequest, was so extensive that a petition to the King in 1724 mentions "two messuages in Crane Court; certain lands and hereditaments in Mablethorpe, Lincolnshire; two houses in Coleman Street . . . and a fee-farm in Sussex" (*ibid.,* vol. i, p. 431 n.).

[53] Eddy, *Critical Study,* ed. cit., p. 163.

satire because of their essential truth to source. The "invention" in Swift's passages usually consists in one of two things: sometimes he neatly combines two real experiments on different subjects—as in the case of the spiders who not only spun silk stockings, but also went one better than the scientists by colouring them naturally; at other times Swift carries a real experiment only one step further—and the added step carries us over the precipice of nonsense.

Two of the "projects" alone seem to have had a literary source. The purposely disgusting experiment of the "most ancient student of the Academy," who attempted to "reduce human excrement to its original food," was based upon Rabelais' "Archasdarpenin." The "Ingenious architect" who built his house, like the bee and spider, by beginning at the roof and working downwards, also had a literary source, though one may suspect that Swift found at least partial authority for the idea in contemporary accounts of architectural experiments.[54] With these two possible exceptions, all of Swift's major experiments may be found in the *Philosophical Transactions* or in more complete works of members of the Royal Society.

The "astronomer who had undertaken to place a sun-dial upon the great weather-cock on the town-house, by adjusting the annual and diurnal motions of the earth and sun, so as to

[54] Eddy points out (*ibid.*, p. 163) the similarity between this passage and one in Tom Brown's *Amusements*. There were however certain architectural experiments not entirely dissimilar, which might well have attracted Swift's attention. Wallis had proposed in 1644 "A Geometrick Flat Floor," working on the problem of how to support a floor over an area wider than the length of the timbers available for joists (*cf.* R. T. Gunther, *Early Science in Oxford*, 1923, vol. i, p. 211). Of a similar nature was the roof of the Sheldonian Theatre, designed by Wren. *Cf.* also "A Bridge without any Pillar under it," "Journal of the Philosophical Society of Oxford," *Phil. Trans.*, 1684, 14, 714; vol. i, p. 594. In an earlier passage on the architecture of Balnibarbi (*Voyage to Laputa*, p. 210), Gulliver finds that a palace may be built in a week "of materials so durable as to last for ever without repairing." Swift at this point may well have remembered a man who discovered, after the event, that he had built his house not of stone but of asbestos (*Phil. Trans.*, Abridged, vol. iv, ii, p. 285).

answer and coincide with all accidental turnings by the wind" was proposing nothing impossible. Such sun-clocks had been invented both in France and in England. Sir Christopher Wren in 1663 had contrived an automatic wind recorder, by annexing a clock to a weather-cock and, by an ingenious combination of a pencil attached to the clock and a paper on a rundle moved by the weather-cock, procured automatic records of the wind.[55] An English correspondent of the Royal Society in 1719, taking exception to the assertion of a Frenchman that "clocks to agree with the Sun's apparent Motion" had been invented first in France, wrote:

[He] supposed that it was a Thing never thought of by any before himself: I shall therefore give this short Account of what I have performed in that Matter myself. . . . [The account follows]. . . . But these Clocks that I then made to agree with the Sun's apparent Time, were done according to the Equation Tables, which I found not to agree very exactly with the Sun's apparent Motion. . . . I made a Table myself by Observation. . . . Since then I have made many of these Clocks.[56]

Others of Swift's experiments follow actual accounts as closely. Among the many remarkable professors of Lagado was "a man born blind, who had several apprentices in his own condition; their employment was to mix colours for painters, which their master taught them to distinguish by feeling and smelling." [57] One might suspect that here Swift was having his fun—as so often—with Newton, particularly

[55] R. T. Gunther, *Early Science in Oxford*, 1923, vol. i, pp. 317–19.

[56] "The Invention of making Clocks to keep Time with the Sun's apparent Motion, asserted by Mr. J. Williamson" (*Phil. Trans.*, 1719, 30, 1080; vol. iv, i, p. 394).

[57] *Voyage to Laputa*, p. 213. Much the same idea—without the mention of a blind man—is used in "Memoirs of Scriblerus," ed. cit., p. 135: "He it was that first found out the Palpability of Colours; and by the delicacy of his Touch, could distinguish the different Vibrations of the heterogeneous Rays of Light." The fact that Swift does not in this earlier work attribute the technique to a blind man may indicate that in the earlier period he was satirizing Newton, while in 1725, in the collected works of Boyle, he found the perfect story for his purposes.

with the corpuscular theory of light, which had been reported in the *Philosophical Transactions*. But the source was more direct; not Newton, but Boyle, was the villain of this piece. Material made to his hand Swift found either in Boyle's "Experiments and Observations upon Colours" or more probably —since Boyle's earlier reports had been made in 1663 and 1664—in the ponderous *Philosophical Works of Robert Boyle,* which had been "abridged, methodized, and disposed under various heads" by Peter Shaw in 1725. The blind professor whom Gulliver saw had in the preceding century been a real blind man, whose case was reported to Boyle by "Dr. Finch, anatomist extraordinary to the great duke of Tuscany." [58] Finch had told Boyle of "a blind man at Maestricht, in the Low Countries, who at certain times could distinguish colours by the touch with his fingers." After several scruples on Boyle's part, he was forced to believe in the account, which he relates in these words:

The name of the man was John Vermaasen, at that time about thirty-three years of age, who, when he was two years old, had the small pox, which render'd him absolutely blind, tho' he is at present an organist in a public choir. The doctor discoursing with him over night, the blind man affirmed, that he could distinguish colours by feeling, but not unless he were fasting; for that any quantity of drink deprived him of that exquisite touch which is requisite to so nice a sensation. Upon this, the doctor provided against the next morning seven pieces of ribbon of these seven colours, black, white, red, blue, green, yellow, and grey; but as for mixed colours, this Vermaasen would not undertake to discern them; tho, if offer'd, he could tell that they were mixed. To discern the colour of the ribbon, he places it betwixt his thumb and his fore-finger, but his most exquisite perception is in his thumb, and much better in the right than in the left.

[58] "Dr. Finch" was evidently Sir John Finch, whose career is described by Archibald Malloch, *Finch and Baines: A Seventeenth Century Friendship* (Cambridge, 1917), and in the *Conway Letters,* edited M. H. Nicolson (New Haven, 1930). Boyle's original paper was written in 1663 and published in 1664. At that time Boyle was prescribing for Lady Anne Conway, sister of Sir John Finch, who received his medical degree at Padua, spent much of his life in Italy and was in close contact with the Grand Duke of Tuscany.

After the man had four or five times told the doctor the several colours, whilst a napkin was tied over his eyes, the doctor observed he twice mistook, for he called the white black, and the red blue; but still before his error, he would lay them by in pairs, saying, that tho' he could easily distinguish them from all others, yet those two pair were not easily distinguishable from one another. Then the doctor desired to know what kind of difference he found in colours by his touch. To which the blind man reply'd, that all the difference he observed, was a greater or less degree of asperity; for, says he, black feels like the points of needles, or some harsh sand, whilst red feels very smooth. . . .[59]

Boyle goes on to point out that before he saw the notes from which the account was taken he had believed that the blind man might have distinguished the colours not by feeling, but by smelling—another point Swift was quick to catch. Boyle's account continues:

for some of the ingredients employ'd by dyers, have different and strong scents, which a very nice nose might distinguish; and this I the rather suspected, because he required that the ribbons he was to judge of, should be offer'd him in the morning fasting; for I have observ'd in setting-dogs, that the feeding of them greatly impairs their scent.

In others of his experiments, Swift has cleverly welded together two or more accounts and made a new combination. The cure "of a small fit of the colic" is of this sort. Here Swift applies to Gulliver a series of experiments Shadwell had already popularized in the *Virtuoso* and implies, in addition, various later experiments performed by members of the Royal Society on the general subject of respiration and artificial respiration.[60] The work of Swammerdam, Hooke, Boyle and others on these subjects had long been familiar; but in addition to this general satire—in which otherwise he might be

[59] *Philosophical Works*, ed. cit., vol. ii, pp. 10–12.

[60] The experiment, satirized by Shadwell, was reported in Sprat's *History of the Royal Society*, ed. cit., p. 232: "By means of a Pair of Bellows, and a certain Pipe thrust into the Wind-pipe of the Creature," artificial respiration was established and its effects observed.

said to follow Shadwell—Swift has suggested something more specific. It is not enough that "a large pair af bellows" should convey air into the intestines or that, when the dog dies from the treatment, artificial respiration should be used to revive it. Swift needed another element, which he found in an account of "An extraordinary Effect of the Cholick," in which Mr. St. Andre had already suggested Swift's principle of "curing that disease by contrary operations":

The Peristaltick Motion of the Intestines is by all Anatomists supposed to be the proper Motion of those Cylindrical Tubes. The use of this Motion is to propel the Chyle into the *Vasa lactea,* and to accelerate the grosser parts of the Aliment downwards, in order to expel them, when all their nutritive Contents are extracted. This Motion, thus established, it naturally seems to follow, that an Inversion of it (call'd for that Reason an Antiperistaltick Motion) should force the Aliments, Bile, Pancreatic Juice, and lastly the Faeces, to ascend towards the Mouth.[61]

The same trick of combining two sources is found in the remarkable experiment of the projector who was able to make silk stockings and gloves from spiders' webs. Swift's projector was found in a room "where the walls and ceiling were all hung round with cobwebs." He lamented "the fatal mistake the world had been so long in of using silk-worms, while we had such plenty of domestic insects, who infinitely excelled the former, because they knew how to weave as well as spin." Emile Pons has suggested that this idea went back to the proposal of a Frenchman; [62] but it has not been noticed that that Frenchman's proposal appeared in the *Philosophical Transactions,* whence it came to Swift's attention. In a paper on "The Silk of Spiders," M. Bon in 1710 first gave an account of various sorts of spiders, which reminds the English reader of the satirical interest in these insects in Shadwell's earlier parody. Shadwell's Sir Nicholas Gimcrack had become intimately acquainted with many kinds of spiders; but M. Bon was con-

---

61 *Phil. Trans.,* 1717, 30, 580; vol. v, i, pp. 270–2.
62 *Gulliver's Travels,* ed. cit., pp. 254 n.

cerned only with two: "*viz.* such as have long legs, and such as have short Ones: The latter of which furnishes the Silk I am going to speak of." M. Bon, however, was aware, as was Sir Nicholas Gimcrack, that spiders "are distinguished by their Colour, some being Black, others Brown, Yellow, Green, White, and others of all these Colours mixed together." Unlike Sir Nicholas Gimcrack, M. Bon was less concerned with species of spiders than with their utilitarian value. He wrote:

The first Thread that they wind is weak, and serves them for no other Use than to make that Sort of Web, in which they catch Flies: The second is much stronger than the first; in this they wrap up their Eggs, and by this means preserve them from the Cold, and secure them from such Insects as would destroy them. These last Threads are wrapt very loosely about their Eggs, and resemble in Form the Bags of Silk-Worms, that have been prepared and loosened between the Fingers, in order to be put upon the Distaff. These Spiders Bags (If I may so call them) are of a grey Colour when they are new, but turn blackish when they have been long exposed to the Air. It is true, one may find several other Spiders Bags of different Colours, and that afford a better Silk, especially those of the Tarantula; but the Scarcity of them would render it very difficult to make Experiments upon them; so that we must confine ourselves to the Bags of such Spiders as are most common, which are the short-legg'd Ones. . . . And by getting together a great many of these Bags, it was that I made this new Silk, which is no-way inferior in Beauty to common Silk. It easily takes all sorts of Colours; and one may as well make large Pieces of it, as the Stockings and Gloves which I have made. . . .[63]

Only one significant difference appears in Swift's account. M. Bon still found it necessary to dye his stockings and gloves in the usual way. But the projector of Lagado had had access to another paper in the *Philosophical Transactions* and was able to produce colours without added expense by a natural method:

He proposed farther that by employing spiders the charge of dyeing silks should be wholly saved, whereof I was fully convinced

[63] *Phil. Trans.*, 1710, 27, 10; vol. v, ii, pp. 21-4.

when he showed me a vast number of flies most beautifully coloured, wherewith he fed his spiders, assuring us that the webs would take a tincture from them; and as he had them of all hues, he hoped to fit everybody's fancy, as soon as he could find proper food for the flies, of certain gums, oils, and other glutinous matter to give a strength and consistence to the threads.

This trick Swift learned from another paper in the *Transactions*, of the very sort that must have delighted his ironic mind. Here Dr. Wall, beginning with a discourse on amber and diamonds, concluded with gum-lac, pismires and artificial and natural dyes, and unconsciously gave rise to experimentation in Lagado:

I don't know in the Animal Kingdom any Thing but Pismires, that affords a Volatile Acid, and in the East-Indies there's a large kind of them that live on the Sap of certain Plants, affording both a Gum and a Colour, which Sap passing through the Body of those Insects and Animals, is by their Acid Spirit converted into an Animal Nature; which is the Reason, that with the Colour extracted from Gum-Lac (which Gum-Lac is nothing else but the Excrements of these Insects or Animals) almost as good, and full as lasting, Colours are made as from Cochineal: I am the more confirmed herein, because I know of an Artificial Way of converting Vegetable Colours into an Animal Nature very much like this, by which the Colours are made much more pleasant and permanent. After the same Manner the remaining Gum, which is an Oleosum, being digested and passing through the Bodies of those Insects or Animals, is by their Volatile Acid converted into a Vegetable Animal Phosphorus or Noctiluca.[64]

The projector whom Gulliver saw "at work to calcine ice into gunpowder" may have been moved by nothing more esoteric than the report of "Haile of so great a Bigness" which "fell at Lisle in Flanders": "One among the rest was observed to contain a dark brown Matter, in the Middle thereof; and being thrown into the Fire, it gave a very great Report." [65]

[64] *Phil. Trans.*, 1708, 26, 69; vol. iv, ii, pp. 275–8.
[65] *Ibid.*, 1693, 17, 858; vol. ii, p. 145. There are several accounts in the *Transactions* of hail and ice, emphasizing the explosive noise of their bursting.

But since Swift's projector had already written a treatise on the "malleability of fire," and since a group of his fellow-projectors, by "condensing the air into a dry intangible substance, by extracting the nitre, and letting the aqueous fluid particles percolate," showed close familiarity with the work of Boyle and his followers, it is more probable that the gunpowder-projector had been studying Boyle's "Experiments and Observations upon Cold," [66] and had been impressed not only by the similarities between heat and cold, but also and more particularly by the long series of experiments on "the expansive forces of congelation" with their recurrent motif of explosion and violence. In all these experiments water is introduced into tubes of various types and allowed to freeze. In all, the tubes break "with a considerable noise and violence" or "the ball of the glass was burst to pieces with a loud report." Occasionally, "the compress'd air flew out with a great noise, and part of the pipe . . . appear'd filled with smoke." Such reports would have been enough for Swift's imagination, even if he had not also read Boyle's paper on "The Mechanical Origins of Heat and Cold" [67] with its discussion of the apparent extravagances of "heating cold liquors with ice." Boyle's persistent interest in both ice and gunpowder is clear enough to the layman, so that Swift need not have entered—though he may—upon the problem of the "effluvia" of both, which Boyle raises in this work.

Of all the experiments of Swift's projectors, none has excited more contemptuous laughter than that of the man who "had been eight years upon a project for extracting sun-beams out of cucumbers, which were to be put into vials hermetically sealed, and let out to warm the air in raw inclement summers." Preposterous as this may seem, it is no more incredible than the other experiments which prove to have scientific sources. Swift merely combined a group of experiments, adding to them little—except the cucumbers! The "cucumber projector" may have been an assiduous student of Grew, Boyle, Hooke and Newton; he may have read a paper by Halley on "The Cir-

[66] *Philosophical Works*, ed. cit., vol. i, pp. 573–730.
[67] *Ibid.*, pp. 550–72.

culation of Watery Vapours," [68] in which many of his ideas were suggested. But it is more likely that he was a follower of Hales, who, working upon principles laid down for him by these predecessors, made the final experiments which were imitated in the Grand Academy of Lagado. Over a period of years Hales had reported to the Royal Society experiments on the respiration of plants and animals, which he welded into a whole in 1727 in his two volumes of *Statical Essays*.[69]

Hales's work also presupposed certain conclusions made by Boyle and Newton on the nature of the "particles of the air," which gave rise to long discussion among men of science. Hales had been particularly impressed by the great quantities of air generated from certain fruits and vegetables, most of all, apples.[70] Swift's projector was familiar not only with the general principles involved in such experiments on plant respiration, but also with a series of experiments reported by Hales upon the effect of sunbeams upon the earth and with the principles by which these sunbeams were alleged to enter into plants:

The impulse of the Sun-beams giving the moisture of the earth a brisk undulating motive, which watery particles, when separated or rarefied by heat, do descend in the form of vapour: And the vigour of warm and confined vapour . . . must be very considerable, so as to penetrate the roots with some vigour. . . . 'Tis therefore probable that the roots of trees and plants are thus, by means of the Sun's

[68] *Phil. Trans.*, 1693, 17, 468; vol. ii, pp. 126–9.

[69] *Statical Essays: Containing Vegetable Staticks: Or, An Account of some Statical Experiments On the Sap in Vegetables.* The first edition appeared in 1727, the second in 1731. The book is, of course, too late to have served as Swift's source; but, as the title-page indicates, the two volumes "incorporate a great Variety of Chymico-Statical Experiments, which were read at several Meetings before the Royal Society." These experiments concerned "the quantities imbibed and perspired by Plants and Trees," in which Hales followed and improved upon Boyle and Hooke. Hales had followed Boyle's experiments, performed with his air-pump and exhausted and unexhausted receivers, upon "Grapes, Plums, Gooseberries, Cherries, Pease" and several other sorts of fruits and grains.

[70] *Vegetable Staticks*, Experiment LXXXVII.

warmth, constantly irrigated with fresh supplies and moisture . . . whence, by the same genial heat, in conjunction with the attraction of the capillary sap vessels, it is carried up thro' the bodies and branches of vegetables, and thence passing into the leaves, it is there most vigorously acted upon, in those thin plates, and put into an undulating motion, by the Sun's warmth, whereby it is most plentifully thrown off, and perspired thro' their surface; whence, as soon as it is disentangled, it mounts with great rapidity in the free air.[71]

Such sunbeams, sinking into the ground, as Hales reported, "for a distance of two feet" and then rising through root and branch of the plant to be "perspired" or "respired," Swift's projector, like Hales, caught in his "hermetically sealed vials." The second step in his experiment needed only the authority of Shadwell's Sir Nicholas Gimcrack, who, motivated by Boyle and Hooke, collected air from all parts of the country so that his guests might choose "Newmarket, Banstead-down, Wiltshire, Bury Air; Norwich Air; what you will"; [72] when Sir Nicholas grew weary of the closeness of London and had "a Mind to take Country Air," he sent for "may be, forty Gallons of Bury Air, shut all my Windows and Doors close, and let it fly into my Chamber." So Swift's projector, having collected the sunbeams, let them out "to warm the air in raw inclement summers."

Practical as was the application of a theory in the case of this botanist, Swift's projectors concerned with "new methods of agriculture" were more practical still. To be sure, they had already, as Gulliver had seen, reduced the fields of Bal-

---

[71] *Ibid.*, pp. 63–6. This was a controversial question in Swift's time, involving as it did questions of the nature of air and of heat. Three years later all these theories of Boyle, Newton and Hales and also that of the Dutch scientist Nieuwentyt were brought together and discussed in a paper in the *Phil. Trans.* by J. T. Desaguliers ("An Attempt to Solve the Phenomenon of the Rise of Vapours," 1729, 36, 6; vol. vi, ii, p. 61).

[72] *The Virtuoso*, in *Dramatic Works of Thomas Shadwell* (4 vols., London, 1720), vol. i, p. 387. The many parallels between Shadwell's work and Swift's would suggest that the *Virtuoso* was one of the important literary sources of the *Voyage to Laputa*. In each instance however Swift has brought Shadwell up to date by material drawn from contemporary science.

nibarbi to desolation and were responsible for those "wholly unaccountable" methods of managing lands which Gulliver had observed. Because of his interest in Ireland, Swift may have noticed particularly a paper "On the Manuring of Lands by Sea-shells in Ireland, by the Archbishop of Dublin," [73] with its suggestions of the great improvements to be wrought in agriculture by this method. Swift may also have noticed the many reports on agriculture in distant countries—particularly Ceylon and China—and the proposals in the *Philosophical Transactions* for carrying over to English soil methods applicable to agriculture in very different climates. Certainly there seem to be reminiscences of such papers in the experiments of the agricultural projectors of Lagado. The professor, for example, who proposed to plough the land by driving six hundred hogs into a field, that they might "root up the whole ground . . . at the same time manuring it with their dung" suggests that he had studied a paper on the "Culture of Tobacco in Zeylan." [74] The custom in Ceylon was this:

They clear a little piece of Ground, in which they sow the Seed of Tobacco, as the Gardeners here sow Parsley and Coleworts; against the Time that this is ready for transplanting, they choose a piece of Ground, which they hedge about; when the Buffalo's begin to chew the Cud, they are put within this Hedge-Ground and let stand until they have done, and this they continue Day and Night, until the Ground be sufficiently dunged.

The "universal artist" who devoted himself to agriculture among his various pursuits, proposed to "sow the land with chaff, wherein he affirmed the true seminal virtue to be contained." The "seminal virtue" of plants had engrossed English botanists since the discovery of Nehemiah Grew that plants possessed sex. Swift needed no source for this interest in Lagado; he could have found it in Grew himself, or in such suggestions as that "Of Manuring of Land by Sea-Sand," in

73 *Phil. Trans.*, 1708, 26, 59; vol. iv, pp. 298–301.
74 *Ibid.*, 1702, 23, 1164; vol. iv, pp. 312–14. (For other possible sources of this passage *cf.* Pons, *Gulliver's Travels*, ed. cit., p. 254, note.)

which the author suggests the mixing "of these Male and Female Salts; for the Sea Salt is too lusty and active of itself." The "Propagation of Vegetables" from this point of view was discussed at length in another paper which, following Grew's discovery "that the Farina . . . doth some way perform the Office of Male Sperm," went on to prove "that this Farina is a Congeries of Seminal Plants, one of which must be convey'd into every Ovum before it can become prolifick." [75]

The most specific proposal of the agriculturists of Lagado Gulliver had observed even before his visit to the Academy. On Munodi's property stood a mill, turned by the current of a river, which had for years proved satisfactory. However, some seven years before, the projectors had come to Munodi

with proposals to destroy this mill, and build another on the side of that mountain, on the long ridge whereof a long canal must be cut for a repository of water, to be conveyed up by pipes and engines to supply the mill; because the wind and air upon a height agitated the water, and thereby made it fitter for motion; and because the water descending down a declivity would turn the mill with half the current of a river whose course is more upon a level.

Behind this sage conclusion of the theorists of the Academy lay a long series of experiments reported to the Royal Society first by Francis Hawksbee, later by James Jurin, who from 1720 until 1727 was the editor of the *Philosophical Transactions*. Beginning with a discussion of the cause of ascent of water in capillary tubes, Jurin continued with a study of the effect on the flow of water at various heights, and finally with two papers, one in English, the other in Latin, "Of the Motion of Running Waters." [76] Here Swift seems to have found,

[75] *Ibid.*, 1708, 26, 142; vol. iv, p. 301; *Ibid.*, 1703, 23, 1474; vol. iv, pp. 305–8.
[76] *Ibid.*, 1718, 30, 748; vol. iv, pp. 435–41; 1722, 32, 179; vol. vi, i, pp. 341–7. In the second, presented to the Society in 1726, Jurin reviews the history of the subject. In the same year J. T. Desaguliers presented (1726, 34, 77; vol. vi, i, pp. 347–50) "An Account of several Experiments concerning the Running of Water in Pipes, as it is retarded by Friction and intermixed Air. . . . With a Description of a new Machine, whereby Pipes may be clear'd of Air, as the Water runs along, without Stand-Pipes, or the Help of any Hand."

couched in the terms of mathematical proof which never failed to amuse and irritate him, a study of the force of running water at various heights, the effect of gravity, the mathematical ratio between "the Altitude of the Water" and the "Motion of the Cataract." Here he may even have found his "Canal"; for Jurin's experiment is in large part a study of the relation between the "Length of the Canal," the "Motion of the Water" and the "Velocity of the Water." To Jurin, to be sure, "Canals" were only a part of laboratory equipment; but by the projectors—and by Swift—the small laboratory model was readily expanded in size and easily translated to the side of a mountain in Balnibarbi, where water flowed (or did not flow, as Munodi learned to his cost) with the same "Force" and "Velocity" as in Jurin's tubes.

Of all the experimenters in the Grand Academy of Lagado, there remains only that "universal artist" who, like Bacon, believed that the end of science was the "benefit and use of man" and who "had been thirty years employing his thoughts for the improvement of human life." So common were "universal artists" in Swift's day that it is perhaps idle to seek to identify the original of a passage which is clearly intended as a satire upon the tendency of many scientists of the time to take all knowledge to be their province. Such a universal artist was the earlier Martin Scriblerus, "this Prodigy of our Age; who may well be called The Philosopher of Ultimate Causes, since by a Sagacity peculiar to himself, he hath discover'd Effects in their very Cause; and without the trivial helps of Experiments, or Observations, hath been the Inventor of most of the modern Systems and Hypotheses." [77] The "universal artist" of the *Voyage to Laputa* was however an experimenter; and his experiments covered many fields. He was at once a specialist on the nature of the air, on petrification, on marble, on agriculture and on the breeding of sheep. Allowing for the obvious exaggeration of the passage, we may suspect that, if Swift intended his thrust at any one man, it was at Robert Boyle, who spoke with authority on all these

[77] "Memoirs of Scriblerus," ed. cit., p. 135.

subjects and who, more perhaps than any other man of his age, had been a pioneer in all fields of investigation. Swift's many other references to Boyle, and his obvious familiarity with the three large volumes in which Dr. Peter Shaw had collected Boyle's works, bear out this theory. But theory it must remain; for Boyle was not alone in his encyclopaedic knowledge and in his tendency to express himself on any and every subject. If vice it was, it was, as Bacon would have said and as Swift seems to imply in this passage, less a vice of the individual than of the age.

VI

"It is highly probable," Swift wrote ironically in the *Voyage to the Houyhnhnms*, "that such travellers who shall hereafter visit the countries described in this work of mine, may, by detecting my errors (if there be any), and adding many new discoveries of their own, justle me out of vogue, and stand in my place, making the world forget that I was ever an author." [78] Swift himself would have been the last to object to the attempts of "later travellers" to recognize the specific sources of his satire. He, who delighted in the setting of riddles, wrote with some regret: "Though the present age may understand well enough the little hints we give, the parallels we draw, and the characters we describe, yet this will all be lost to the next." Yet he added more hopefully: "However, if these papers should happen to live till our grandchildren are men, I hope they may have curiosity enough to consult annals and compare dates, in order to find out." [79] Letters exchanged between Swift and his contemporaries offer evidence that Swift's own age was quick to catch the implications in the scientific por-

[78] *Gulliver's Travels*, ed. cit., p. 349.

[79] *Prose Works*, edited Temple Scott, vol. ix, p. 110. Sir Charles Firth, one of the most acute of the commentators on *Gulliver's Travels*, said (*The Political Significance of Gulliver's Travels*, ed. cit., p. 1): "A critic who seeks to explain the political significance of *Gulliver's Travels* may be guilty of too much ingenuity, but he cannot fairly be charged with exaggerated curiosity. He is searching for a secret which Swift tells us is hidden there, and endeavouring to solve riddles which were intended to exercise his wits."

tions of the voyage. Ten days after the publication of *Gulliver's Travels*, Gay and Pope wrote Swift a joint letter, in which they said that there was general agreement by the politicians that the work was "free from particular reflections, but that the satire on general societies of men is too severe." [80] They added: "Not but that we now and then meet with people of greater perspicuity, who are in search for particular applications in every leaf, and it is highly probable we shall have keys published to give light into Gulliver's design." Erasmus Lewis complained that he wanted such a key.[81] Dr. Arbuthnot, recognizing the satire upon his colleagues in the Royal Society, wrote critically to Swift: "I tell you freely, the part of the projectors is the least brilliant"; [82] Gay and Pope reported to Swift that Arbuthnot had said "it is ten thousand pities he had not known it, he could have added such abundance of things upon every subject." [83] To the joint letter Swift replied, still pretending to preserve his anonymity, reporting other criticisms which had come to him. He added a sentence which may well be significant in connection with the *Voyage to Laputa:* "I read the book over, and in the second volume observe several passages, which appear to be patched and altered, and the style of a different sort, unless I am much mistaken." [84] Various explanations may be suggested for that self-criticism. In the light of the evidence which has been offered here, is it not possible that Swift intended an apology for the haste with which the scientific portions of the *Voyage to Laputa* were completed, after his return from Ireland? [85]

---

[80] *Works of Alexander Pope . . . with Introduction and Notes,* by Rev. Whitwell Elwin, London, 1871, vol. viii, p. 88.

[81] *Correspondence,* ed. by F. Elrington Ball, London, 1912, vol. iii, p. 357.

[82] *Prose Works of Jonathan Swift,* edited Temple Scott, vol. viii, p. xvi.

[83] *Works of Pope,* ed. cit., vol. vii, p. 89.

[84] *Ibid.,* vol. vii, pp. 91–2.

[85] Cf. *Remarks on the Life and Writings of Dr. Jonathan Swift . . . In a Series of Letters from John Earl of Orrery To his Son,* London, 1752, p. 99: "He seems to have finished his voyage to Laputa in a careless hurrying manner, which makes me almost think that sometimes he was tired of his work, and attempted to run through it as fast as he could."

Whatever the artistic inferiorities of the *Voyage to Laputa*, Swift has left to posterity in these chapters a record of the greatness and the limitations of his time. No age will be a "Century of Genius" that does not also appear to its coevals a century of absurdities. Perhaps the final word on this adventure of Gulliver may best be said, not by posterity, but by one of Swift's contemporaries, John, Earl of Cork and Orrery, who wrote to his son:

However wild the descriptions of . . . the manners, and various projects of the philosophers of Lagado may appear, yet it is a real picture embellished with much latent wit and humour. It is a satire upon those astronomers and mathematicians, who have so entirely dedicated their time to the planets, that they have been careless of their family and country, and have been chiefly anxious, about the economy and welfare of the upper worlds. But if we consider Swift's romance in a serious light, we shall find him of opinion, that those determinations in philosophy, which at present seem to the most knowing men to be perfectly well founded and understood, are in reality unsettled, or uncertain, and may perhaps some ages hence be as much decried, as the axioms of Aristotle are at this day. Sir Isaac Newton and his notions may hereafter be out of fashion. There is a kind of mode in philosophy, as well as in other things. . . .[86]

[86] *Ibid.*, p. 97.

VI. The Microscope and
English Imagination

LONG before the microscope gave man proof of the existence of a world of minutiae beyond the vision of the human eye, poetic fancy and fairy legend had imagined such a world. The "small" has had a perennial fascination more powerful—because more intelligible—than the "great." Certainly as early as the Bible, men found pleasure in studying tiny insects and drawing lessons from them. Early "spectacles" and simple magnifying glasses intensified that interest. "Regio-Montanus his Fly" was admired "before his Eagle," by others than Sir Thomas Browne.[1] Queen Mab, "the fairies' midwife," was

---

[1] "Indeed what Reason may not go to School to the Wisdom of Bees, Ants, and Spiders? what wise hand teacheth them to do what Reason cannot teach us? Ruder heads stand amazed at those prodigious pieces of Nature, Whales, Elephants, Dromedaries, and Camels; these, I confess, are the Colossus and majestick pieces of her hand: but in these narrow Engines there is more curious Mathematicks; and the Civility of these little Citizens more neatly sets forth the Wisdom of their Maker. Who admires not Regio-Montanus his Fly beyond his Eagle?" *Religio Medici*, in *The Works of Sir Thomas Browne*, edited Geoffrey Keynes (1928), I, 20.

> In shape no bigger than an agate-stone . . .
> Drawn with a team of little atomies . . .
> Her waggoner a small grey-coated gnat,
> Not half so big as a round little worm
> Prick'd from the lazy finger of a maid.

In addition to perennial interest in the minute, various other important ideas led pre-historic thinkers to logical theories presupposing a realm of existence beyond the eye. On the one hand, early theories of contagion—*contagium vivum* or *contagium animatum*—suggested the existence of minute "carriers" of malaria and other epidemic diseases. Then too there was the neo-Platonic conception of Nature as a *plenum formarum,* a theory which led to the belief that there must be a minutely graded hierarchy, a *continuum* of forms from highest to lowest—the "principle of plenitude," as Professor Lovejoy has called it. As faith and reason postulated the existence of a series of orders above man, reason at least anticipated *a priori* the possibility of the existence of minute orders below the human, animal, and vegetable kingdoms. The proof of the actual existence of such a world of minute life came as no surprise or shock. Human reason had anticipated it. Instruments merely offered proof of its existence.

We shall not therefore find, as a result of the invention of the microscope, any such immediate stimulation of the imagination as study of the telescope has disclosed. There is nothing here to correspond to that night in 1610 when Galileo saw through heaven, no one book to parallel the effect of the *Sidereus Nuncius.* Yet if the influence of the microscope on imagination was not so startling as that of the telescope, it is not less interesting and pervasive. Both because the microscope was more easily used by the amateur and because the world of minutiae was more readily comprehensible, the smaller instrument had an even more popular appeal. The heavens might declare the glory of God, but the handiwork showed by the firmament remained beyond the comprehension of many. The perfection of a tiny insect, the actual sense-evidence for "millions of Eels in a sawcer of Vinegar"—these man's eyes could see and his mind comprehend.

From the scientist the microscope passed to the amateur and the layman, stimulating both serious and satiric themes in literature. Modern sophisticated readers may smile at the excitement caused in a Tyrolese village by the discovery among a dead man's effects of a strange instrument in which appeared "a devil . . . shut up in a glass," which caused the mayor and village fathers to refuse Christian burial to its owner. The modern reader is aware, as the Tyrolese discovered only after terrified controversy, that "the devil proved to be a bristly and hairy flea." [2] Modern "Men of Wit," like their seventeenth-century predecessors, may laugh at the interest and enthusiasm of early amateur-scientists and of "learned ladies," in lice and fleas and the twisting "little animals" of Leeuwenhoek, may think naive the "Philosophical Girl" who dissected her dove and studied the circulation of blood in the tail of a fish while her lover watched the circulation in "this pretty Neck." But behind the humor of the early popular enthusiasm for the microscope lay something more serious. Addison smiled over the dissection of a beau's head or coquette's heart, but he also read microscopical implications into a passage of Locke's. Swift satirized the new science with characteristic vehemence, but *Gulliver's Travels* would not be what it is had Swift not looked through a microscope—perhaps the one he bought for Stella—and felt the fascination and repulsion of grossly magnified nature. The microscope brought new themes to literature, new conceptions that were to influence ethics, aesthetics, metaphysics and ideas of God and of man. If it caused charming ladies to shriek at the magnified appearance of their delicate skin, it led to a fuller understanding of this "best of all possible worlds," and to a new awareness of man, "The glory, jest, and riddle of the World."

I

As in the history of the telescope, with which it is intimately concerned, the beginnings of the microscope are lost in an-

[2] This is one of several microscopical tales told by Gaspar Schott in his *Magia Universalis Naturae et Artis,* 1657.

tiquity and in legend.[3] Optical instruments were probably unknown in the classical world, but legend persistently finds them in the Middle Ages, whether in such fantasies as Kipling imagined in *The Eye of Allah,* in mysterious references of Roger Bacon, or in early accounts of spectacles "for the help of poor blind old men" in Italy about 1300. Such spectacles were ancestors of the first magnifying-glasses and of the simple lenses known and used before the invention of the compound microscope. Like the telescope, the compound microscope had a dual origin in Holland and in Italy. The name of the instrument—*microscopio* or *microscopium*—seems first to have been used by Giovanni Faber, in a letter to Prince Federigo Cosi on April 13, 1625.[4] But the microscope gathered to itself no such poetic nomenclature as did the "optick glasse" in England. *Microscope* it was, and *microscope* it remained for the most part in both scientific and popular terminology.

The English inventor of the microscope, as of the telescope, was probably Thomas Digges, though there are no suggestions in the descriptions left by Digges or by his son of important microscopical discovery. In a way, Francis Bacon may be considered the spiritual ancestor of microscopy in England. "Neither the naked hand nor the understanding left to itself can effect much," Bacon said. "It is by instruments and helps that the work is done, which are as much wanted for the understanding as for the hand."[5] His insistence upon the

[3] Excellent general bibliographies on the subject may be found in Clifford Dobell, *Antony van Leeuwenhoek and his "Little Animals,"* 1932, pp. 398–422, and in A. N. Disney, C. F. Hill, W. E. Watson Baker, *Origin and Development of the Microscope,* London, 1928, pp. 116–152, 284–297.

[4] Faber (quoted Disney, pp. 98–9) was speaking of Galileo's microscope. He said: "I also mention his new *occhiale* to look at small things and call it *microscope.*" See also Edward Rosen, *The Naming of the Telescope,* pp. 23–4, 65–6. One of the few variant terms for the new instrument was the "Smicroscopium" of Athanasius Kircher; "multiplying glass" was occasionally used.

[5] *Novum Organum,* Aphorism II. Bacon, of course, also stressed the dangers of trusting too implicitly in instruments. Cf. Aphorism I: "For the sense by itself is a thing infirm and erring; neither can instruments for

need of "instruments and helps" had much to do with the fact that one chapter in the history of the Royal Society occurred among a group of Oxford men who met in the rooms of an expert in glass-grinding. There were other elements, too, in the Baconian teachings that led to interest in the microscope among his followers: his belief that the function of the scientist was "to examine and dissect the nature of the very world itself"; the emphasis in his method upon *counting, weighing, measuring;* his belief that "there are still laid up in the womb of nature many secrets of excellent use, having no affinity or parallelism with anything that is now known," with his suggestion that some of these might be discovered in "the remoter and more hidden part of nature"; his insistence that science must deal with things "mean or even filthy, things which (as Pliny says) must be introduced with an apology—such things, no less than the most splendid and costly, must be admitted into natural history. . . . For whatever deserves to exist deserves also to be known . . . and things mean and splendid exist alike." [6]

Whether or not Francis Bacon knew the microscope, his Father of Salomon's House told visitors to the New Atlantis about such an instrument. "Wee procure meanes of Seeing Obiects a-farr off: As in the Heaven and Remote Places," he declared. The sentence is not remarkable for 1627; but at least one later phrase is unusual in a period so early: "We have also Helps for the Sight, farre above Spectacles and Glasses in use. Wee have also Glasses and Meanes to see the small and Minute Bodies, perfectly and distinctly: As the Shapes and Colours of Small Flies and Wormes, Graines and Flawes in Gemmes which cannot otherwise be seen, *Observations in Urine and Bloud not otherwise to be seen.*"

The chief stimulus to the development of the microscope in England was not native but came from the continent,

---

enlarging or sharpening the senses do much." Much Restoration satire on microscopical observation had to do with amateurs who leaped to absurd conclusions on the basis of casual observation.

6 *Novum Organum,* Aphorism CXX.

particularly from Descartes' *Dioptrique*. The most interesting work of this period to the layman was the *Century of Microscopic Observations* [7] of Pierre Borel, in which Borel listed one hundred observations, showing that the microscope was being used in the study of plants, animals, insects, as well as for casual observation. By 1660, the compound microscope was recognized as an instrument of scientific value, while great strides had been made in the development of simple lenses. Indeed, it may justly be said that the year 1660 marks the beginning of the importance of the microscope in England. Before that time, while interesting observations had been made, important workers in the field had devoted themselves rather to the *principle* of lenses than to significant observation. The quarter-century following saw the establishment of modern botany, physiology, microbiology, protozoology, and bacteriology. In England, during the period 1660–85, the microscope developed from a mere novelty into an important adjunct to the investigations of the Royal Society. Then occurred a period of enthusiasm for its possibilities, followed by an equal period of waning interest and of comparative disuse of the instrument among scientists, accompanied by a growing enthusiasm for the microscope among laymen.

An early English reference to microscopical observation may be found in the preface written in 1634 by Sir Theodore Turquet de Mayerne to Thomas Mouffet's *Theatre of Insects*.[8] Mayerne wrote, commenting upon studies Mouffet had

[7] The *Centuria Observationum Microcospiarum* constituted the third part of Borel's *De Vero Telescopii Inventoro*, 1655. It was published separately in 1656. The spelling *microcospiarum* is consistent throughout.

[8] *The Theatre of Insects: or Lesser Living Creatures*, written, according to the preface, toward the end of Elizabeth's reign, has sometimes been considered the earlier English microscopic book. Closer examination, however, shows that Mouffet's observations were made with the naked eye. Mayerne, in the preface dated 1634, says that various suggestions had been made for its publication, but it does not seem to have appeared until Edward Topsell published it, with other works, in 1658 in his *History of Four Footed Beasts, Serpents, and Insects*.

made with the naked eye: "If you take lenticular Glasses of crystal (for though you have Lynx his eyes, these are necessary in searching after Atoms) . . . you will doubtless enter upon a large field of Philosophy concerning three Kingdoms of the universal spirit (the Vegetable, Animal, and Mineral) equally penetrating, replenishing, and governing." The first important English work on microscopical observation, however, was the *Experimental Philosophy* of Henry Power,[9] published in 1664. Power's long *Preface* is full of information and suggestions. He surveys the history of "Dioptrical Glasses," and points proudly to the fact that they are a *modern* invention; clearly on the side of the "new" in the long warfare between *tradition* and *progress*, he uses the microscope as proof of the superiority of the "moderns" over the "ancients," and of "art" over "nature." As the telescope had made man aware that he dwells in no narrow and straitened world, monarch of a single globe, so the microscope, he suggested, would show him a new universe of another sort, and men would find themselves, Power prophesies, "but middle proportionals (as it were) 'twixt the greatest and smallest Bodies in Nature, which two Extremes lye equally beyond the reach of human sensation."

II

Yet though the volumes of Mouffet and Power were important, they were only forerunners of the great movement. The history of the development of the microscope in England may be read in the life and work of one man—Dr. Robert Hooke, Curator of the Royal Society.[10] To the Royal Society, indeed, the instrument came to be "Hooke's microscope." It

[9] *Experimental Philosophy, in Three Books: Containing New Experiments Microscopical, Mercurical, Magnetical. With some Deductions, and Probable Hypotheses, raised from them in Avouchment and Illustration of the new famous Atomical Hypothesis, By Henry Power, D. of Physick*, London, 1663-4.

[10] The quotations from Hooke, unless otherwise indicated, may be found under the date in the *Philosophical Transactions of the Royal Society* or, more conveniently, in R. T. Gunther, *Life and Works of Robert Hooke* in *Early Science in Oxford*, VI (1930).

was he who first introduced microscopical observations as a part of the routine of Society meetings; during the period of the greatest scientific interest in the instrument, he was instructed to prepare for each meeting of the Royal Society at least one observation.

Hooke's first formally reported observations were given at the meeting of April 22, 1663, when he "brought in two microscopical observations, one of leeches in vinegar; the other of a bluish mold upon a mouldy piece of leather." On May 6 he reported "a microscopical observation of a female gnat"; on May 20, three observations, on the head of an ant, on the point of a needle, on "a strange fly like a gnat." Simple enough "observations" from the point of view of modern science but the microscope was still a novelty and comparatively few people could use it. The following week Hooke was "charged to look upon sage with a microscope, and to observe, whether there lurked any little spiders in the cavities of the leaves, that might make them noxious." Hooke's next observations led him into a field later to prove full of startling possibilities. On June 17, 1663, "Dr. Wilkins mentioned, that Dr. Croune had, in the blood of a dog dissected by him, found abundance of little insects. Mr. Hooke was desired to take notice thereof, and to make frequent observations with a microscope on the blood of several animals." Perhaps Hooke, rather than Leeuwenhoek, might have been remembered as the father of bacteriology, had his microscope been powerful enough. At this time he was trying to develop the technique of observation, for on July 3, 1663, he wrote to Boyle, "I have made a microscope object glass so small, that I was fain to use a magnifying glass to look upon it, but it did not succeed so well as I hoped." [11]

At this point Hooke's observations were interrupted for a time by notice that the royal patron of the Royal Society was about to make a visit to the assembly. As Curator, Hooke was responsible for the equipment and demonstrations on that

[11] Gunther, *op. cit.*, p. 131.

occasion. Among the preparations the Society ordered, "Mr. Hooke was charged to show his microscopical observations in a handsome book to be provided by him for that purpose." In this volume, designed for the casual interest of Charles II, undoubtedly appeared the originals of some of those exquisite drawings that make Hooke's *Micrographia* a delight to scientist, artist, and layman.

The visit of the King successfully concluded, the Royal Society settled down to reports of experiments and observations. Since the interest of the society had been attracted to the possibility of the microscopical examination of liquids, Hooke was next directed, on August 17, 1663, to observe with his microscope "some rain-water with a great number of little insects in it . . . and to draw the picture of them. The operator was directed to keep those insects, in order to see, whether they would turn into any other kind." So intent was Hooke upon the "insects" of whose existence he already knew that again he failed to perceive the "animalcules" Leeuwenhoek's keener observation detected. Or perhaps, to do Hooke justice, his glasses were still at fault; for in the period that follows, he mentions more than once his difficulties in procuring the proper equipment from "Mr. Reeves," the most expert glass-grinder then in England.

In view of the difficulties Hooke encountered, the plates of his *Micrographia* are the more remarkable. On November 3, 1664, the Royal Society having examined the manuscript of that volume, ordered that "the President be desired to sign a licence for the printing of Mr. Hooke's microscopical book," but added a word of warning:

That Mr. Hooke give notice in the dedication of that work to the Society, that though they have licensed it, yet they own no theory, nor will be thought to do so; and that the several hypotheses and theories laid down by him therein, are not delivered as certainties, but as conjectures; and that he intends not at all to obtrude or expose them to the world as the opinion of the Society.[12]

12 *Ibid.*, p. 219.

The Royal Society need have had no fears, for with the exception of the *Preface*, the *Micrographia* is singularly free from "hypotheses and theories." Indeed, in its clear and accurate observation, its objective reports of observation, its tentative conclusions, it is a model of Baconian science, as in its close observation of facts, accumulation of data, and unwillingness to suggest hypotheses except upon the evidence of an accumulation of instances, it is a monument to the Baconian method.

From the time of the publication of the *Micrographia*, the Royal Society considered microscopical observation and reports an important part of every meeting, and realized the potential value of the instrument. When in 1667 another distinguished—if fantastic—visitor sent word of a proposed visit to the Royal Society, Hooke was immediately ordered to provide not only a book of observations, but a "good microscope" to be used upon that occasion. At the visit of "Mad Madge" of Newcastle, which moved Pepys to irony, the microscopical demonstrations were presided over by Hooke and Robert Boyle. From this time on, information about new instruments was brought to the meetings. So rapid was advance in the construction and technique of the microscope by Hooke and Sir Christopher Wren that Sprat in his *History of the Royal Society* in 1667 used the instrument as one of the most important proofs of the "advancement" of *modern* thought.

Even before this time circumstances had led members of the Royal Society to question whether, by use of the microscope, man might find the secret of the plague. The Great Plague of 1665 still seemed to the majority of its victims a judgment of God upon a perverse generation, but more thoughtful minds questioned whether plagues were not rather of human origin and capable of human solution. Writing to Boyle after the Royal Society had temporarily disbanded, Hooke suggested that the plague was spread by contagion or infection, had much to say of climate and temperature, and commented upon the flies and insects of the season.[13] His letter of July 8, 1665 suggests one reason that the next chapter in the history of the microscope, though continental in origin, was English in development. On October 15, 1677, Hooke read to the Royal Society a letter from Antony van Leeuwenhoek in which English scientists heard for the first time almost incredible accounts of the earliest discoveries in microbiology.

Antony van Leeuwenhoek [14] is one of the most remarkable figures in the history of science. A man entirely without scientific training, in no sense a philosopher, not even a learned man, as the century understood learning, he wrote no book or treatise, left no "works." His early observations were a result of mere curiosity. His first microscope was only a simple magnifying-glass, which in his trade he had used to examine the linens he officially inspected. He had, however, a genius for observation, and uncanny eyesight, as any modern scientist who attempts to use even his later microscopes must realize. The first of the observations that were to transform medicine and biology was made in September, 1674, when, passing by a nearby lake, Leeuwenhoek took up a pail of water in which he first observed microscopically the "little animals." "And the motion of most of these animalcules in the water," he concludes, "was so swift, and so various, upwards, downwards, and round about, that 'twas wonderful to see; and I judge that some of these little creatures were above a thousand times smaller than the smallest ones I have ever yet seen, upon the rind of cheese, in wheaten flour, mould and the like." Leeuwenhoek's second observation was upon rain-water, in September, 1675. "I discovered," he noted, "living creatures in rain which had stood but a few days in a new tub, that was painted blue within . . . more than ten thousand times smaller than the animalcules . . . called by the name of

13 *Ibid.*, pp. 247–9.
14 In 1932 Clifford Dobell published his compendious *Antony van Leeuwenhoek and his "Little Animals,"* which so far transcends all previous studies that I limit my references to that volume.

Water-flea or Water-louse." In April, 1676, he prepared infusions of pepper-water, in which he later observed "little creatures," reports of which precipitated experiments in the Royal Society and satire among English dramatists. Not himself a draughtsman, he employed an assistant for his drawings; one of the charming features of his reports is the exclamations of his draughtsman, whom the master taught to see a new world of life. In making drawings of a flea, Leeuwenhoek tells us, the artist constantly interrupted his observations to exclaim: "Lieve God wat sijnder al doneren in soo een kleyn schepsel!" [15]

Even before this time Leeuwenhoek had begun another sort of observation that was to have important consequences. He first studied intestinal protozoa, which he discovered in the bile of various animals and in human bacteria. He tested his own mouth and teeth, discovering, as he had suspected, the presence of "little animals" in the human body. No letter is more interesting than the "Letter on the Protozoa" written October 9, 1676. In this earliest account of the observation of protozoa, Leeuwenhoek wrote: "I now saw very plainly that these were little eels or worms, lying all huddled up together and wriggling; just as if you saw, with the naked eye, a whole tubful of very little eels and water, with the little eels, a-squirming among one another: and the whole water seemed to be alive with these multifarious animalcules. This was for me, among all the marvels that I have discovered in nature, the most marvellous of all."

Leeuwenhoek was an observer, not a philosopher. We do not find in him the reaches of imagination we shall discover in others. Perhaps because of his objectivity and the simplicity of his style, his occasional ponderings are the more telling and significant. He stopped frequently to marvel at "such perfection in this tiny creature." He commented on the "curious mathematics" revealed by a world of minute life. He was amazed "to see such a diversity of structures" in "these delight-

[15] Dobell, p. 147 n.

some and wondrous little creatures which quite escape the bare eye." Throughout his reports we find the deepening conception of a greater power—call it Nature or God—as the "secret world" is more and more fully displayed before him. "Seeing these wondrous dispensations of Nature," he wrote, "whereby these little creatures are created so that they may live and continue their kind, our thoughts must be abashed."

Finding no interest in his ideas among continental thinkers, Leeuwenhoek began correspondence with the Royal Society through the secretary, Henry Oldenburg, a correspondence interrupted by the secretary's death. The first letter Hooke brought before the Society on October 15, 1677 commented upon Oldenburg's death, and went on to describe Leeuwenhoek's recent observations on pepper-water. The Society immediately directed Hooke to make a microscope like Leeuwenhoek's and repeat his observations. The first English attempts were unsuccessful. For a time it seemed that the sceptics and satirists were justified. But the patient Hooke persisted, and on November 15, 1677 reported that he too had seen "great numbers of exceedingly small animals swimming to and fro. They appeared of the bigness of a mite through a glass, that magnified about a hundred thousand times in bulk." Not only had he himself observed the animalcules, but he offered an impressive list of witnesses: "Mr. Henshaw, Sir Christopher Wren, Sir John Hoskyns, Sir Jonas Moore, Dr. Mapletoft, Mr. Hill, Dr. Croone, Dr. Grew, Mr. Aubrey, and divers others"—the first Englishmen to see the new world of life beyond the human eye.

The immediate interest in Leeuwenhoek's discoveries is shown in the second visit of Charles II to the Society. "His majesty having been acquainted with it, was desirous to see them and very well pleased with the observation," noted Hooke in a letter to Leeuwenhoek. Hooke and other members of the Society set themselves with new energy to the development of glasses. From 1677 until 1682 microscopical experiments of the Royal Society were chiefly concerned with attempting to improve on Dutch observations.[16] We find in addition further observations on anatomy, particularly on the muscles of lobsters, crabs, shrimps, that brought ironic comment from more than one satirist. For at least five years, interest in bacteriology and protozoölogy ran high.

And then, almost as abruptly as the microscopical enthusiasm had begun in England, it waned. This may be attributed, in part, to the temporary eclipse of the Royal Society, thanks to the satirists. Added to this was the fact that the Society found itself in the rôle of imitator rather than originator; in spite of all their labors, they had produced no results so significant as those of Leeuwenhoek. Recognizing his limitations in this new field of observation, Hooke turned back to earlier work in mechanics and physics. From 1682 until 1691 the minutes of the Royal Society contain no important microscopical observations reported by him. There was a general feeling among scientists, indeed, that this field had been exhausted. The Royal Society turned to newer problems.

That Hooke did not forget his earlier enthusiasm may be seen in a discourse he delivered to the Royal Society on telescopes and microscopes in February, 1691/2—perhaps the most competent survey of the subject in the seventeenth century.[17] Unlike many of his contemporaries, he did not make the mistake of attributing the development of optical instruments to Galileo alone, but discussed the work of Roger Bacon, Porta, Metius, Digges, and others. He realized that a period of stagnation had set in:

But tho' there has been some life left in the grinders of glasses, yet, the warmth of those, that should have used them, has grown cool; and little of new discoveries hath been made by them. . . . Much the same hath been the fate of microscopes, as to their invention, improvements, use, neglect and slighting, which are now reduced almost to a single votary, which is Mr. Leeuwenhoek; besides whom, I hear of none that make any other use of that instrument, but for

16 See the minutes of the Society for December 13, 1677 and March 14, 1677-8.

17 The paper is given in Gunther, pp. 735 ff.

diversion and pastime, and that by reason it is become a portable instrument, and easy to be carried in one's pocket.

The microscope, Hooke believed, had passed from scientists to become the plaything of the public. Only Leeuwenhoek remained, a lonely figure without followers, the butt of satire, a fantastic who propounded absurdities. Hooke was correct in his analysis of the situation. For a time the microscope ceased to be an important scientific instrument and became the toy of the aristocracy—most of all, of the "ladies." Unlike the telescope, it was comparatively cheap; its use required little professional skill; it was easily understood by the layman. As it ceased temporarily to affect the history of science, it became an important influence in the history of literature.

III

Characteristically, the first suggestion of popular interest in the microscope occurs in the diary of Samuel Pepys. On February 13, 1663/4, Pepys and Creed paid a visit to "Reeves, the perspective glass maker," where they saw "very excellent microscopes, which did discover a louse or a mite most perfectly and largely." On July 26 Pepys again visited Reeves, this time to select a microscope for himself which, he says, cost five pounds, ten shillings, "a great price," he adds, "but a most curious bauble it is, and he says, as good, nay, the best he knows in England, and he makes the best in the world." During the period between his ordering and receiving the microscope Pepys' mind was running on the subject. On August 7, as he walked home, he met Mr. Spong with whom he had talked "of many ingenuous things, musique, and at last of glasses." Mr. Spong had made a microscope for himself through which he had discovered "that the wings of a moth is made just as the feathers of the wing of a bird, and that most plainly and certainly." Pepys bought Power's *Experimental Philosophy,* just off the press. On the evening Reeves delivered his microscope, Pepys read in Power's book "to enable me a little how to use and what to expect from my glasse." Like most amateurs, he encountered difficulties, for he noted the next evening that he and his wife had made their first observations "with great pleasure, but with great difficulty before we could come to the manner of seeing anything by my microscope." At last they succeeded "though not so much as I expect when I come to understand it better." Almost as soon as it was published, Pepys secured from his bookseller "Hooke's book of microscopy, a most excellent piece, and of which I am very proud." His interest in the microscope was still keen a year and a half later, when on July 29, 1666, Mr. Spong and Reeves dined with him, and after dinner proceeded "to our business of my microscope." During that year Pepys grew interested also in the telescope, thanks to Mr. Reeves, who on August 7, brought with him a "twelve-foote glasse" which they set up on the "top of the house." In spite of some difficulties Pepys was at last able to make telescopic observations also, and jotted down in his diary comments on various phenomena he observed.

Pepys and Mr. Spong were not alone in their interest in the new glasses. Cowley in his *Ode* mentioned the microscope as one of the triumphs of the Royal Society:

> Nature's great Works no Distance can obscure,
> No Smallness her near Objects can secure.
>> You've taught the curious Sight to press
>> Into the privatest Recess
> Of her imperceptible Littleness . . .
>> You've learned to read her smallest Hand,
> And well begun her deepest Sense to understand.

As in the case of the telescope, poets and dramatists turned to the new instrument for figures of speech, for poetic imagery, for themes of satire and irony. Samuel Butler had his fun with new styles of poetry:

> He that would understand what you have writ
> Must read it through a Microscop of wit;
> For evry line is Drawn so curious there
> He must have more then eies that reads it cleare.

He laughed at

> . . . one, who for his Excellence
> In height'ning Words and shad'wing Sense,
> And magnifying all he writ
> With curious microscopick Wit,
> Was magnify'd himself no less
> In home and foreign Colleges.

In a satiric portrait of the learned man, he suggested:

> What Nature had to human eies denyd
> He with the optiques of his minde discry'd.

The interest of the Royal Society in such observations as Hooke's on vinegar, mould, and gnats, seemed at first only a matter for laughter. Butler, whose favorite theme for satire was always the *virtuosi*, found new material in microscopical observations. While *The Elephant in the Moon* was primarily directed against contemporary enthusiasm over the telescope, the microscope did not escape. Among the characters was

> . . . one, whose Task was to determin
> And solve th' Appearances of Vermin;
> Wh' had made profound Discoveries
> In Frogs, and Toads, and Rats, and Mice.[18]

In *Hudibras,* Butler had ridiculed men who concerned themselves with the pulse of a "dappled louse," the jump of a flea, or such problems as

> How many different Specieses
> Of Maggots breed in rotten Cheeses;
> And which are next of kin to those
> Engender'd in a Chandler's nose;
> Or those not seen, but understood,
> That live in Vinegar and Wood.[19]

[18] "Poetry," in *Satires and Miscellaneous Poetry and Prose,* edited René Lamar, 1928, p. 245; "The Elephant in the Moon," *ibid.,* p. 7; "Learning," *ibid.,* p. 157; "The Elephant in the Moon," *ibid.,* p. 12.

[19] *Hudibras,* Part II, canto 3.

Andrew Marvell was equally amused at the absurd observations of Hooke and the fervor of "Mad Madge" when he instructed a painter:

> With Hook then, through the microscope, take aim
> Where, like the new Controller, all men laugh
> To see a tall Lowse brandish the white Staff. . . .
> Paint then again her Highness to the life,
> Philosopher beyond Newcastle's wife.[20]

The most complete satire on the observation of what Bacon called "mean or even filthy things" occurs in Shadwell's *Virtuoso*.[21] Many of the audience who laughed with Shadwell had, like Pepys, read Power and Hooke and played with the new toy. Although Shadwell satirized the telescopic interests of the Royal Society, his Sir Nicholas Gimcrack was pre-eminently a microscope-enthusiast. He had spent two thousand pounds on microscopes to "find out the Nature of Eels in Vinegar, Mites in a Cheese, and the Blue of Plums," and had "broken his brains about the nature of Maggots and has studi'd these twenty years to find out the several sorts of Spiders." Like many gentlemen—and some ladies—of the period, he had his own "Elaboratory . . . a spacious Room, where all his Instruments and fine Knick-Knacks are." There was no microscopic suggestion made by Hooke Sir Nicholas did not pursue. He had studied ants, "dissected their Eggs upon the object plate of a Microscope," and had "watch'd whole days and nights" their peculiarities and habits. He was versed in flies, and learned in the "fabrick and structure" of spiders. "No man upon the face of the earth," Sir Formal

[20] "Last Instructions to a Painter," in *Poems and Letters of Andrew Marvell*, edited H. M. Margoliouth, 1927, pp. 141–2. Marvell suggests the technique of the microscope in "Upon Appleton House" (*ibid.*, I. 73) in the stanza ending

> Such Fleas, ere they approach the Eye
> In Multiplying-Glasses lye.

[21] *The Virtuoso*, in *The Dramatick Works of Thomas Shadwell*, London, 1691. The drama was played and published in 1676. Acts III and IV are particularly concerned with scientific satire.

declared, "is so well seen in the Nature of Ants, Flies, Humble-bees, Ear-wigs, Mille-pedes, Hogs-Lice, Maggots, Mites in a Cheese, Tadpoles, Worms, Neuts, Spiders." Gimcrack knew the latest scientific terminology, and overwhelmed his hearers with such terms as "Fluidity, Orbiculation, Fixation, Anguliza-tion, Crystallization, Germination, Ebullition, Vegetation, Plantanimation." Like Hooke and Borel, he had detected "Eels" in liquids, and offered to show "Millions in a Sawcer of Vinegar: they resemble other Eels, save in their Motion, which in others is side-ways, but in them upwards and downwards, thus, and very slow." He had studied the "Blue upon Plumbs," which he discovered to be living creatures, by ob-servation "upon a Wall-Plumb (with my most exquisite Glasses, which cost me several thousands of pounds)." As a result of his observations, Sir Nicholas concluded that "the whole Air is full of living Creatures, a thousand times less visible than those living Creatures, mistaken for Motes in the Sun." Like many members of the Royal Society and other amateurs of the day, Sir Nicholas was fascinated by the possibilities of micro-scopical dissection. He had invited his guests "to his House, to see a Cock-Lobster dissected," but when the demonstration proved impossible because the fish-monger failed to deliver the lobster, the virtuoso-host proposed another treat: "After Dinner we will have a Lecture concerning the Nature of In-sects, and will survey my Microscopes, Telescopes, Thermom-eters, Barometers, Pneumatick Engines, Stentrophonical Tubes, and the like."

Shadwell's *Virtuoso* was a prelude to an outburst of mingled irony and enthusiasm for such themes. On the one hand, we hear persistent laughter at "collectors" [22] whose "rare speci-mens" of mean and insignificant objects seemed to the age fantastical and absurd; on the other, there is a new interest in anatomical dissection under the microscope, impressing even those who laughed at the devotees.

[22] Walter Houghton has studied the *virtuoso* as "collector," as well as in many other aspects in "The English Virtuoso in the Seventeenth Century," *Journal of the History of Ideas*, III (1942), 51–73, 190–219.

When in 1698 Ned Ward made his tour of London, described in *The London Spy*,[23] he visited Gresham College, the home of the Royal Society, which had become in his mind "Maggot-Monger's Hall." On his visit to the "Elaboratory-Keepers Apartment, for a sight of his rarities," he found "Philosophical Toys," and "rarities," whose worthlessness and inconsequence so impressed him that he abruptly "removed from Maggot-Monger's Hall," and went instead to visit an asylum for the admittedly insane. His general impression of the follies of the scientist is seen in his description of the "character of a virtuoso," who, he remarks satirically, "has a Closet of Curiosities outdoes Gresham Colledge." In the same year, William King took occasion to satirize such "journeys" as that of Samuel Sorbière in 1664, and Dr. Martin Lister's "journey to Paris." In "Mr. Shuttleworth" King discovered another *virtuoso* who was a collector:

He shew'd me a Thousand other Rarities, as the Skin of a Cap Ass, many very excellent Land Snails, a Freshwater Mussel from Chatham; a thin Oyster; a very large Wood Frog, with the Extremity of the Toes webbed. He shew'd me some Papers of Swammerdam, in which were some small Treatises, or rather some Figures only of the Tadpole. Again, Figures relating to the Natural History of a certain Day-Butterfly, and of some considerable number of Snails, as well Naked as Fluviatile.

The Ninth Dialogue of King's *Dialogues of the Dead*[24] contributes still another chapter to the "Battle of the Books." "Moderno" is discovered "such a hideous Figure and . . . so dirty, that no Gentleman would come near you." He has not been thrown from his horse—the only accident that could excuse his appearance, but has been in a ditch, hunting tadpoles.

23 *The London Spy, Compleat in Eighteen Parts, by Ned Ward. With an introduction by Ralph Straus*, London, 1924, pp. 59 ff. This is reprinted from the London edition of 1700.

24 "Modern Learning," *Dialogues of the Dead*, in *Miscellanies in Prose and Verse*, pp. 324–338. The dialogue is a satire on William Wotton's *Reflections on Ancient and Modern Learning*, quotations from which appear in the speeches of "Moderno."

"There has been more true Experience in Natural Philosophy,"
he declares, "gather'd out of Ditches in this latter Century,
than Pliny and Aristotle were Masters of both together." The
"Winter Sports" of this Modern are "Rat-catching, Mouse-
fleying, Crevice-searching for Spiders, Cricket-dissecting and
the like," but in the spring he takes to the fields, seeking
"Maggots, Flies, Gnats, Buzzes, Chaffers, Humble-Bees, Wasps,
Grasshoppers, and in a good year Caterpillars in abundance."
He refers easily to "Goedartius and Swammerdam." He boasts
of an acquaintance who has studied "all those Excrescences
and Swell-Twigs" and another who "has made many Observations upon Insects that live, and are carry'd about upon the
Bodies of other insects." He rebukes Indifferentio for his
amusement, insisting that "all these excellent Men do highly
deserve Commendation for these seemingly useless Labours,
and the more, since they run the hazard of being laugh'd at
by Men of Wit."

"Men of Wit" continued to laugh, and the laughter grew
louder as amateur observation and collection became more
popular. "Mr. Willis of St. Mary Hall, Oxon." produced "A
Comical Panegyrick on that familiar Animal by the Vulgar
call'd a Louse," [25] addressing that popular insect in mock-
heroics:

> Tremendous Louse, who can withstand thy Power,
> Since Fear, at first, taught Mortals to adore?
> What mighty Disproportion do we see
> In Adam's Glory, when compar'd with thee? . . .
>
> Who can thy Power describe, thy Glories scan,
> Thou Lord of Nature, since thou'rt Lord of Man?
> In these we may thy wondrous value see,
> The World was made for Man, and Man for thee.

Tom Brown, who was responsible for this panegyric, had
more to say about the new vogue. " 'Tis reckoned a great part

[25] The two stanzas quoted are given by Richmond Bond, *English Burlesque
Poetry*, 1932, p. 254. The poem was published in *The Works of Mr. Thomas
Brown in Prose and Verse*, London, 1707, pp. 18 ff.

of Learning," he remarked in his *Table Talk*,[26] "to know the Names of Things. We have some Virtuosos that can nicely distinguish the meanest Mosses, yet know nothing of their Vertue and Efficacy, which is just all one, as if a Foreigner should come to London, and get all the Signs of Cheapside, and Cornhill, and not trouble himself to know anything of the Government of the City." Later William Shenstone wrote another poem "To the Virtuosi." [27]

> Hail, curious Wights! to whom so fair
>   The form of mortal flies is!
> Who deem those grubs beyond compare
>   Which common sense despises. . .
>
> Let Flavia's eyes more deeply warm,
>   Nor thus your hearts determine,
> To slight Dame Nature's fairest form,
>   And sigh for Nature's vermin.

The classic expression of interest in the microscopic louse and the flea is Swift's stanza, which every age has quoted and misquoted:

> So, naturalists observe, a flea
> Has smaller fleas that on him prey;
> And these have smaller still to bite 'em,
> And so proceed *ad infinitum*.[28]

Perhaps the most extended satire upon collectors of microscopic specimens is the "Will of a Virtuoso" in an essay in *The Tatler* for August 26, 1710. The will was Nicholas Gimcrack's, who left to members of his family many "rarities" he had examined through his microscope, among them, "a dried cockatrice . . . three crocodile's eggs . . . my last year's collection of grasshoppers . . . my rat's testicles," together with

[26] In *Works*, p. 40.

[27] *Levities: or Pieces of Humour*, in *Poetical Works of William Shenstone*, edited George Gilfillan, 1854, pp. 74–6, stanzas 1 and 10.

[28] "On Poetry," in *The Poetical Works of Swift*, edited John Mitford, 1880, II. 74.

"all my flowers, plants, minerals, mosses, shells, pebbles, fossils, beetles, butterflies, caterpillars, grasshoppers and vermin, not above specified." In other moods, Addison showed his enthusiasm for microscopical observation, but on this occasion he launched his light barbs at the collectors, "a sort of learned men who are wholly employed in gathering together the refuse of nature . . . able to discover the sex of a cockle, or describe the generation of a mite." It is, he says, "the mark of a little genius to be wholly conversant among insects, reptiles, animalcules, and those trifling rarities that furnish out the apartment of a virtuoso. . . . Whatever appears trivial or obscene in the common notions of the world, looks grave and philosophical in the eyes of a virtuoso."

The "Tatler" might laugh with other "Men of Wit," but the collecting instinct of the *virtuosi* was strong enough to withstand irony. "Mr. Shuttleworths" of the day continued to collect specimens of lowly insects which they examined with passionate eagerness under their microscopes. In the Land of Giants, Lemuel Gulliver, after his heroic fight with wasps, was careful to preserve four of their stings; "upon my return to England," he noted, "I gave three of them to Gresham College, and kept the fourth for myself." Passion for insects was satirized in Mrs. Centlivre's *Bold Stroke for a Wife* by Mistress Lovely, who warned:

Ay, study your Country's Good, Mr. Periwinkle, and not her Insects.—Rid you of your homebred Monsters, before you fetch any from abroad—I dare swear you have Maggots enough in your own Brain to stock all the Virtuoso's in Europe with Butterflies.[29]

The interest in anatomy that had been growing steadily during the period of the Renaissance received new stimulus from the invention of the microscope. Not only was it now possible for the human eye to observe what had before been mere matter of conjecture, but careful study began to prove unrecognized similarities between human anatomy and that of lower forms of life, which was to have far-reaching effect

[29] *Works of Mrs. Centlivre*, 1761, III. 252.

upon medical diagnosis and treatment. The microscope really founded the science of physiology, an important chapter in which dates from the development of a microscope through which it was possible to observe the circulation of the blood.

Notices in almanacs of the period suggest the popular interest in such microscopes. An advertisement in *News from the Stars* [30] in 1698 announced: "At the Old Archimedes and Spectacles in Ludgate Street, liveth John Marshall. . . . He hath invented a new Microscope fit for all Objects, and particularly for shewing the Circulation of the Blood in little Fishes." The 1700 number of the almanac of *Merlinus Anglicus, Junior,*[31] standard for many years, carried an advertisement of John Yarwell, "approved of by the Royal Society," stating that he had invented a "New Double-Microscope, whereby may be seen the Circulation of the Blood in small Fishes." In the long history of the warfare between the "humanities" and the "sciences," there is no more ironic chapter than the quarrel between Meric Casaubon, son of the humanist, and Pierre Gassendi, one of the greatest French scientists.[32] Gassendi had declared that his microscopical study of the circulation of blood in a gnat had first led him to understand the psychological effects of physiological processes. Observing an insect through the microscope and seeing the result of anger upon circulation, he learned to control his temper. Casaubon replied in effect that it had taken no microscope to teach him simple lessons he had learned at his mother's knee and from the Holy Bible.

Long before the invention of the microscope, "physiological poetry" had been common. But the microscope gave new direction to this kind of writing, introducing ideas as novel as they were interesting. Matthew Prior's *Alma: or the Progress*

[30] *News from the Stars: Or, An Ephemeris for the Year 1698 . . . By William Andrews,* London, 1698. Like most almanacs it is unpaginated.

[31] *Merlinus Anglicus Junior: Or The Starry Messenger, 1700* (by Henry Colby).

[32] The controversy is discussed at length by Richard F. Jones, *Ancients and Moderns,* St. Louis, 1936.

of the Mind and Blackmore's *Creation* were to a later period what Fletcher's *Purple Island* had been to the earlier. Prior's poem is filled with anatomical description of the new sort. Man may now, he suggested,

> to an inch compute the station
> 'Twixt judgment and imagination.

He described the brain in detail, drawing technical terminology from scientific authorities, and treated in turn the optic nerves, the nerves of fingers and toes, the ear drums, the sense of taste. Descartes' theory that the pineal gland was the seat of the soul had been a subject for the satirists ever since Descartes proposed it. Microscopical study gave it new life, as may be seen in *The Guardian* for April 21, 1713, which satirically suggests the enthusiasm of the scientists. Problems of respiration were given new impetus by Richard Lower's discovery that the bright color of the blood is due to the absorption of air in its passage through the lungs, by Boyle's experimentation proving that air is necessary for life, by Swammerdam's work on respiration, and by Hooke's experiments in artificial respiration. Various reports reached the Royal Society at the same time of experiments in blood transfusion, which afforded convenient new material for satirical situations in drama and poetry.[33]

The current scientific patter of "ladies" and "gentlemen" is reflected in Mrs. Centlivre's *Bold Stroke for a Wife,* when "Mrs. Lovely," speaking to the pseudo-scientist "Periwinkle," cries: "What, would you anatomize me?" to which another character promptly replies: "Ay, ay, Madam, he would dissect you . . . or, pore over you through a Microscope, to see how your Blood circulates from the Crown of your head to the

[33] A summary of some of the chief experiments on respiration may be found in the abridgement of the *Philosophical Transactions and Collections, to the End of the Year MDCCC,* I. 49, II. 214–235. Claude Lloyd has discussed some experiments on respiration and blood-transfusion in "Shadwell and the Virtuosi," *Publications of the Modern Language Association of America,* XLIV (1929). 472 ff.

Sole of your Foot." [34] The interest aroused in the popular mind by technical anatomical matters was not always satirical. A long section of Blackmore's *Creation* is devoted to close description of parts of the human body, the process of the circulation of the blood, and a versification of the main facts of anatomy, particularly those that had been proved or discovered by the microscope. Blackmore was a physician and therefore the detail was familiar to him, but it was intelligible also to his readers, who had followed the anatomical discoveries reported to the Royal Society. With Blackmore they could say:

> The Living fabrick now in pieces take,
> Of every part due observation make.

Under the microscope Blackmore dissected the human body, showing "the various bones so wisely wrought," the "nice strings of tended membrances," the "arterial pipes in order laid," "Organs perplext, and clues of twining veins." He studied the function of "th' elastic spirits," which "now begin to work the wondrous frame," examined "the beating heart . . . by turns dilated, and by turns comprest," followed the course of the arteries and veins, "the crimson jets with force elastic thrown" and the "fine complicated clues of nervous thread." We trace other processes, follow other functions, until, having observed "Nature's many intricacies,"

> To sublimer spheres of knowledge [we] rise
> By manag'd fire, and late-invented eyes.

To a modern reader, Blackmore's account seems parody, but to its own generation it was a serious treatment of the wonders of man's body, never before so clearly understood.

The most familiar satire upon physiological observations is the dissection of "The Beau's Head" and "The Coquette's Heart" reported by *The Spectator* in January 1711–12.[35] Here is the "assembly of virtuosos," with their "many curious ob-

---

[34] *Works of Mrs. Centlivre*, ed. cit., III. 251.
[35] *Spectator*, Numbers 275, 281.

servations . . . lately made in the anatomy of the human body . . . by the help of very fine glasses." The technique of the *Spectator* papers was that of microscopical anatomists. The "beau's head" is "opened . . . with a great deal of nicety." At first glance it seems like any other head, "but upon applying our glasses to it, we made a very odd discovery, namely, that what we looked upon as brains, were not such in reality, but an heap of strange materials wound up in that shape and texture, and packed together with wonderful art in the several cavities of the skull." Gravely the "operator" and the audience study the structure of the brain under the microscope; they find the pineal gland "encompassed with a kind of horny substance, cut into a thousand little faces or mirrors, which were imperceptible to the naked eye." They examine "a large antrum or cavity in the sinciput," and "several little roads or canals running from the ear into the brain," which they trace "out through their several passages." They study the *musculi amatorii* of the eye, and "the elevator, or the muscle which turns the eye towards heaven," and note their observations on "the skull, the face, and indeed the whole outward shape and figure of the head." Stripped of its satire, we have here the outline of a report to the Royal Society. At the end, the scientists report that the specimen will be kept "in a great repository of dissections" where others may study at leisure this strange aberration from the normal.

In the "Coquette's Heart," *The Spectator* satirizes another popular aspect of the new microscopic physiology—the effect of the circulation of the blood on the action of the heart. The operator warns the audience of the difficulty of such dissection "by reason of the many labyrinths and recesses" in the specimen before him. He calls attention first to the *pericardium;* "by the help of our glasses, [we] discerned in it millions of little scars." Study of the "reddish liquor" leads one of the number to report an experiment he has made by substituting such liquid for mercury in his thermometer. The dissection proceeds, until the operator lays bare the heart itself. The observers note the temperature, and comment on the slipperiness of the outward surface. They study the nerves, and discover a connection between the heart and the eye, though none with the brain. Not content with dissection alone, they use the specimen for an experiment with heat—with dire results that destroy the organ so that it has not been preserved for future investigation.

The *Tatler, Spectator,* and *Guardian* appealed to feminine no less than to masculine readers. In the light of many references in all these periodicals to the new science, it is no surprise to find that one chapter in the development of popular interest in the microscope appeared in the "ladies' periodicals," and that satire upon the "learned lady" broadened to include the "scientific girl."

IV

Perhaps the interest of the "Philosophical Girl" in microscopical science may be dated from the visit of the Duchess of Newcastle to the Royal Society on May 30, 1667. "Mad Madge" she may have been, but the very audacity of a woman forcing a visit to that august body stirred to envy less daring ladies, who had heard from their lords of observations reported at Gresham College. Mistress Pepys, it will be remembered, joined her husband in his evenings over the microscope. The position of the intelligent woman of the Restoration, so far as scientific matters were concerned, is symbolized in Milton's Eve. Eve remained during the Angel's account of the War in Heaven, remained even during the story of the Creation, which, in the light of Adam's original question, might well have been considered a metaphysical account of ultimate origins. When the conversation turned to astronomical theories, however, Eve left the two men to their discussion. Yet as Milton pointed out, she did not go because she was not interested nor because she could not follow such abstractions:

> Yet went she not as not with such discourse
> Delighted, or not capable her ear
> Of what was high. . . .

She had every intention of hearing the discourse later. A good wife, a canny woman (and a "lady" before she was a "scholar"), she knew the pleasure in store for Adam in explaining to her the latest theories. An affectionate wife, she anticipated the "grateful digressions" and "conjugal caresses" with which these would be sweetened to make hard problems just a little easier for a woman who knew her place in the Scale of Being. Yet even Eve might have envied the audacity of the Duchess of Newcastle, who invited herself to the Royal Society, and for whom Hooke and Boyle prepared microscopical demonstrations.

Before her visit to Gresham College, the Duchess of Newcastle had published a philosophical work in which she showed curiosity about contemporary scientific *theories,* though little knowledge of experimentation. Increasing interest may be detected in her *Experimental Philosophy,*[36] written in 1666, shortly before her visit to the Royal Society, and reprinted the following year. In the conclusion the noble authoress pleaded for more contemplation and less experimentation in science, perhaps because of her realization that she could not keep up with newer methods and had better stick to older.

Satires on the "learned lady" in England are so common, as a result of Molière's *Les Femmes Savantes,* that it is no surprise to find a group that treat less the "blue-stocking" than the lady of scientific interests. In the early period of satire on the *virtuosi,* the "scientist" was usually masculine; the ladies, when they appeared, joined in the laughter against him. With their lovers, they poked fun at his enthusiasm, or like Shadwell's Miranda and Clarinda, inveighed against a besotted guardian who wasted his wealth upon microscopes rather than upon their doweries. In one of his plays, Shadwell introduced a "Virtuosa," learned in medicinal lore, who spoke with authority of *Flos Unguentorum, Paracelsian,* and *Green-salve,* and

[36] The earlier work, *Philosophical and Physical Opinion,* was published in 1653. In *Experimental Philosophy,* section 3 deals with the microscope, section 34 with the telescope. There are other allusions to the instruments in the Duchess of Newcastle's *CCXI Sociable Letters,* 1664.

who had compounded an *Album Grecum*.[37] Yet this was little more than a satiric portrait of the conventional "learned lady," who loved high-sounding language and boasted of her Latinity.

The immediate ancestress of the *virtuosa* in England was the Marchioness in Fontenelle's *Conversations on a Plurality of Worlds.* Published in 1686, and translated several times into English, this volume marks the real emergence of the "Philoso-phress," who no longer felt it necessary to subordinate herself to her lord in her interest in the new science. Fontenelle's Lady was an even more provocative figure than the Duchess of Newcastle to the English ladies of the upper classes. Here was no extravagant character at whom Pepys and Evelyn laughed, but a woman of breeding, culture, intelligence and beauty, admired and respected by the Philosopher who strolled with her between hedges of clipped roses, and like Adam mingled with his technical analyses of the Cartesian system, "grateful digressions." In addition to the astronomical dis-cussion, which is the main theme of the volume, the Phi-losopher described to his pupil the world of the microscope as Borel, Hooke, Swammerdam, and Leeuwenhoek had shown it. He too had gazed through a microscope and been amazed at another new world. That amazement he transmitted to the Marchioness. As there are inhabitants in the planets, he told her, so there is life in another world beyond man's sight:

You must not think, that we see all the living Creatures that inhabit the Earth. For there are as many several species and kinds of Animals invisible, as there are visible. We see distinctly from the Elephant to the Mite; there our sight is bounded, and there are in-finite numbers of living Creatures lesser than a Mite, to whom, a Mite is as big in proportion as an Elephant is to it. The late inven-tion of Glasses call'd Microscopes, have discover'd thousands of small living Creatures. . . .[38]

[37] *The Sullen Lovers,* 1669.
[38] *A Discovery of New Worlds, from the French, Made English by Mrs. A. Behn,* London, 1688, p. 91.

184

The Philosopher told his pupil about experiments of Leeuwenhoek and Hooke, but with none of the ridicule of Shadwell in the *Virtuoso:*

How lately have our Virtuoso's found out the Pepper Worms, which in the least drop of Water appear like so many Dolphins, sporting in the Ocean; nay, they tell you that the sharpness of Vinegar consists in the fierceness of the little Animals that bite you by the Tongue: not to name the blue on Plums, and twenty Experiments of the like nature. . . . They have discovered that several, even of the most solid Bodies, are nothing but an immense swarm of imperceptible Animals.

Appreciating the bewilderment of his pupil, the Philosopher plucked a leaf from the hedge. "Do but consider this little leaf," he suggested:

Why it is a great World, of a vast extent, what Mountains, what Abysses are there in it? the insects of one side, know no more of their fellow Creatures on t'other side, than you and I can tell what they are now doing at the Antipodes. . . . In the hardest Stones, for Example, in Marble, there are an infinity of Worms, which fill up the vacuums, and feed upon the substance of the Stone; fancy then millions of living Creatures to subsist many years on a grain of Sand. . . . You will find the Earth swarms with inhabitants.[39]

Such passages as these, with Fontenelle's romantic telescopic pictures of the world in the moon, turned the ladies of England as of France to an eager if superficial interest in the discoveries of the new instruments. Of the two, the microscope naturally made the greater appeal to ladies. Easily obtainable at a not prohibitive price, it could readily be used by amateurs, and its discoveries were immediate and intelligible. In a short time it became the ladies' toy. As contemporary advertisements indicate, grinders of glass (like cigarette manufacturers in our time) found in women a new buying public. Exquisite glasses were available for them in specially prepared cases,

[39] *A Plurality of Worlds, Translated into English by Mr. Glanvill,* 1702, pp. 89–90.

which might easily be carried in place of snuff-boxes. *Virtuosae* of the day delighted in the new fad. Swift wrote to Stella on November 15, 1710: [40]

I called at Ludgate for Dingley's glasses, and shall have them in a day or two; and I doubt not it will cost me thirty shillings for a microscope, but not without Stella's permission; for I remember she is a *virtuoso.* Shall I buy it or no? 'Tis not the great bulky ones, nor the common little ones, to impale a louse (saving your presence) upon a needle's point; but of a more exact sort, and clearer to the sight, with all its equipage in a little trunk that you may carry in your pocket. Tell me, sirrah, shall I buy it or not for you?

So the scientific interests of the ladies flourished. They picked up the vocabulary of the new science. They were as interested as their husbands in accounts of dissection, and followed the implications of microscopy as eagerly as men. "Calphurnia" in William King's "Affectation of the Learned Lady" [41] was "so visited in a Morning by the Virtuosi" that she "had little time to talk with my Milliner, Dresser, Mantua-maker, and such illiterate people." She lost touch with the normal world of woman: "My Day for the Ladies was but once a Fortnight, but every Day for the Virtuosi."

Not only did the "Philosophical Girl" of those days talk with *virtuosi;* as an experimenter, she began to find a place in drama. Two examples, over a quarter of a century apart, will suggest the vogue. In 1693 Thomas Wright adapted Molière's *Les Femmes Savantes* in *The Female Virtuosos,* the title of which suggests changing interest in the "learned lady." While the situations in the two plays are somewhat similar, Wright's "Lovewitt" is a dabbler in science who has been making experiments. In 1726 James Miller satirized in his *Humours of Oxford* "Lady Science," "a formidable pretender to Learning and Philosophy" who knew all the jargon of pseudo-science, had much to say of instruments, and upbraided her less ad-

---

[40] *Journal to Stella,* London, 1922, p. 55. There is a later entry on the same subject under December 22, 1710.

[41] *Dialogues of the Dead,* in *Miscellanies in Prose and Verse,* pp. 303–310.

vanced niece for lack of enthusiasm for Oxford—"this Galaxy, as I may say, from whence are transplanted all the glorious Luminaries that irradiate our Hemisphere." [42]

Of all the treatments of the "Virtuosa" in this period, none is more amusing than *The Bassett-Table* of Mrs. Susannah Centlivre.[43] Here the "virtuosa" has become the heroine. While there is still laughter at her expense, there is also a contagious quality in her enthusiasm for science, which explains the patience of her lover, "Lovely, an Ensign," who calls her with impatient tenderness "that little She-Philosopher," sends her specimens for dissection instead of elegant trifles, and has "stood whole Hours to hear her assert, that Fire cannot burn, nor Water drown, nor Pain afflict, and Forty ridiculous Systems." At her first entrance Valeria, "a Philosophical Girl" comes "running" in pursuit of "the finest Insect for Dissection, a huge Flesh Fly, which Mr. Lovely sent me just now, and opening the Box to try the Experiment, away it flew." She is impatient with such conventional women as "Lady Reveller," who shrieks with horror at hearing that Valeria has dissected her dove "to see whether it is true that doves lack gall." When Lady Reveller suggests sarcastically that Valeria bestow her fortune in "founding a College for the Study of Philosophy, where none but Women should be admitted," Valeria replies that she would gladly do so, were her money under her control. Valeria has turned her feminine dressingroom into a laboratory where she performs "Experiments on Frogs, Fish, and Flies." For the first time a Restoration "bedroom scene" is laid in a laboratory, where Valeria is discovered "with Books upon a Table, a Microscope, putting a Fish upon it, several animals lying by." When Lovely enters, she calls him to look through her glass "and see how the Blood circulates in the Tail of this Fish." In spite of Lovely's soft suggestion that he

---

42 *Humours of Oxford, a Comedy . . . by a Gentleman of Wadham College*, 1730, p. 12. The scientific satire in the play is less concerned with the microscope than with the telescope. This is true of a number of satires on the *virtuosa* which I have not mentioned.

43 In *The Works of Mrs. Centlivre*, 1761, Vol. I.

finds the circulation of blood "prettier in this fair Neck," she shows him her curiosities, among them "the Lumbricus, Laetus, or Foescie, as Hippocrates calls it, or vulgarly in English, the Tape-Worm." The humor of the situation is further enhanced by a variation upon the customary "concealment" of Restoration comedy. When her father's step is heard and discovery seems inevitable, Valeria hides her lover in a tub in which she usually keeps fish, then coolly warns her father not to touch it, since it contains "a Bear's young cub that I have bought for Dissection." In spite of the satire, Mrs. Centlivre is sympathetic toward her young and attractive *virtuosa*. In the popular literature of the microscope, there is no more amusing "modern" speech than the response of this "Philosophical Girl" when her lover begs her to elope with him: "What! and leave my Microscope?"

As Fontenelle in France aroused the interest and curiosity of ladies of France and England in telescopic and microscopic science, so a half-century later, Francesco Algarotti in Italy stimulated their interest in Newtonianism. The main teachings of *Il Newtonianismo per le dame* [44] lie beyond our present scope, but the book reflects so clearly the current interest of learned ladies in the microscope that one section may detain us. As Aphra Behn had translated the earlier work for the benefit of her countrywomen, Elizabeth Carter translated the Italian. The book was appropriately dedicated to Fontenelle, whose *Plurality of Worlds* "first softened the savage Nature of Philosophy, and called it from the solitary Closets and Libraries of the Learned, to introduce it into the Circles and Toilets of Ladies." Explaining to his Lady the limitations of

---

[44] *Il Newtonianismo per le dame: ovvero, dialoghi sopra la luce e i colori*, Napoli, 1737. The work was translated into French as *Le Newtonianisme pour les Dames, ou Entretiens sur la Lumière, sur les Couleurs, et sur l'Attraction . . . par M. Duperron de Castera*, Paris, 1738. The English translation by Elizabeth Carter appeared as *Sir Isaac Newton's Philosophy Explain'd*, London, 1739; another edition, London, 1742; another Glasgow, 1765. I have discussed this popularization of Newton further in *Newton Demands the Muse*, Princeton, 1946.

the senses unless they are assisted by instruments, Algarotti illustrated his point by showing her her own lovely skin as it appears through the microscope, a demonstration that brought from the horrified Countess the prompt reply, "Nature has done us a great favour in not making our senses too refined." In the Third Dialogue, he described in more detail the principles of optical instruments, and continued:

Microscopes have made us in Reality see an infinite number of Animals of which we had not the least Knowledge before, in Things which were not looked upon as very proper to afford them a Habitation. Not to say any Thing of the Discoveries of Anatomy and Natural History, which we owe to these Glasses: Aromatic Infusions, a Drop of Vinegar, are peopled by so prodigious a Number of little Animals that Switzerland and China would appear empty and uninhabited when compared to them. . . . It is an amazing Thing to reflect upon the Minuteness, Art, and Curiosity of the Joynts, Bones, Muscles, Tendons, and Nerves necessary to perform the swift Motions of the smallest microscopical Animals. These Discoveries shew us in how little a Compass all Art and Curiosity may be comprised, even in a Body less than a small Grain of Sand, and yet as compleat, as exquisitely formed, and as finely adorned as that of the largest Animal. Their Multiplicity is no less surprising than their extreme Smallness.

Eighteenth-century ladies were supposed to be *au courant* with the literature of the microscope in several languages. In one passage Algarotti developed Pope's "Why has not man a microscopic eye?"; in another, we find an analogy drawn from *Gulliver's Travels,* in still another a paraphrase of Leibniz. To the Italian Countess as to most English women, the "Pigmy Worlds" of the microscope were more intelligible than the "immense and Gigantic Scene of Vortices or Suns diffused over the whole Universe" which the telescope had shown. The Philosopher teaches simple lessons of relativity: "The little has its Beauties as well as the great, or rather . . . there is no great or little, but with regard to ourselves." It is the microscope, he says, "and that infinite Number of Pigmy Worlds discover'd by it, which has rectified our Ideas of great or little." Having caught her interest, the Philosopher deftly leads his pupil, as apt as the French Marchioness, to the subject of "infinitely small Quantities which has made so great a Noise in the learned World," and thus to Newton's theories of "the same order of Infinities in the Succession of Time, as there are in Extension," and still more technical matters. A potent instrument, the microscope, by means of which philosophers instructed ladies of France, Italy, and England in principles of the infinitesimal calculus!

Fontenelle was the precursor of a long line of less gifted writers who, like Milton's Adam, undertook to soften the "savage Nature of Philosophy" for weaker intellects. The eighteenth century produced a succession of works on "popular science"—science for ladies, science for children, or, as a writer of the eighteenth century more elegantly phrased it, science "made clear even to the meanest capacity." Such volumes deserve a study in themselves, with their extension of the Baconian idea that "Experiments of Fruit" and "Experiments of Light" were not to be the portion of a few "intellectuals," but when the "right method is followed," and the Sense and Reason are supported by "instruments and helps," they might be shared by all mankind. To be sure, Bacon had not anticipated a part played in the advancement of science by women, yet he would have seen in the great popular interest in these matters only a broader interpretation of his "Great Instauration." He would have agreed with George Adams, who in his *Essays for the Microscope*,[45] urged microscopical study as a substitute for pursuits of the gaming table "to which numbers unhappily sacrifice their health and beauty." Such employment, he pointed out, is "infinitely superior to a rational mind":

Investigations of this kind particularly recommend themselves to the attention of the ladies, as being congenial with that refinement of taste and sentiment, and that pure and placid consistency of conduct which so eminently distinguished and adorn those of this happy

[45] George Adams, *Essays for the Microscope*, 1798, p. 666.

isle. To the honour of several ladies of eminence be it recorded, that they are proficients in the study of the various branches of natural history, and many others are making considerable progress in this pleasing science; than which, none can possess a greater tendency to sweeten the hours of solitude and anxiety.

A climax of feminine enthusiasm for the microscope may be found in Eliza Haywood's *Female Spectator*.[46] The essays were supposedly the production of a Club composed of four women, the "Female Spectator" herself who "added to a genius tolerably extensive . . . an education more liberal than is ordinarily allowed to persons of my sex"; "Mira, a lady descended from a family to which wit seems hereditary"; "a widow of quality, who not having buried her vivacity in the tomb of her lord, continued to make one in all the modish diversions of the times"; and Euphrosine, "the daughter of a wealthy merchant, charming as an angel, but endued with so many accomplishments, that to those who know her truly, her beauty is the least distinguished part of her." Among the "modish diversions of the times" reflected in the characteristic interests of this cross-section of polite society, scientific observation held an important place. Expeditions of the ladies to the country to study "Nature" began as the result of a long letter from "Philo-Naturae" supposedly written from the Inner Temple on April 27, 1745, urging the study of Natural Philosophy. He would not "advise them to fill their heads with the propositions of an Aldrovanus, a Malebranche, or a Newton . . . the ideas of those great men are not suited to every capacity," but suggested *observation* and *collection,* in which ladies might hope to equal their masculine contemporaries. Let them go out into the fields; "We cannot walk, or throw our eyes abroad, without seeing ten thousand and ten thousand living creatures, all curious in their kind." Let them not despise "mean or even filthy things." "Even those worms which appear most despicable in our eyes, if

46 *The Female Spectator* was issued from April 1744 to May 1746. Quotations here are from *The Female Spectator, by Mrs. Eliza Haywood. The Seventh Edition.* London, 1771. The chief section on science occurs III. 124 ff.

examined into, will excite our admiration"; "those flying insects, which are most pleasing to the eye, spring from such as but a few days past crawled upon the earth"; "the beauty of the gaudy butterfly . . . rises from the groveling caterpillar." But even such curiosities which may be seen with the naked eye "are infinitely short of those beyond it." Since Nature has not given man's sight the power of discerning minute creation, art has supplied the deficiency. The microscope will afford ladies new wonder, showing the color, form, exquisite workmanship of insects, opening before them a new infinity of nature. "The glasses which afford us so much satisfaction," he added, "are now as portable as a snuff-box, and I am surprized the ladies do not make more use of them in the little excursions they make." Such study will result not only in "pleasing amusement" and "matter for agreeable conversation," but, suggests "Philo-Naturae" wickedly, "subjects of astonishment" so that even "those of the greatest volubility will . . . want words and be silent before the wonders of the universe." As they walked in the country "in little troops" they were to carry their microscopes; "what a pretty emulation there would be among them, to make fresh discoveries." Even the ladies might find new species, and the Royal Society would be indebted to every fair Columbus for the discovery of a new world. "To have their names set down . . . in the memoirs and transactions of that learned body, would be . . . a far greater addition to their charms than the reputation of having been the first in the mode." "Philo-Naturae" stressed still another appeal of the "new science": "All this pleasure, this honour, this even deathless fame may be acquir'd *without the least trouble or study.*"

"Philo-Naturae's" advice had its effect. The Female Spectator, Mira, Euphrosine, and the Widow went forth into "Nature" equipped with observing minds and artificial eyes. It was unfortunate that the English climate proved unpropitious, forcing them "to stay, for the most part, within doors, and pass our hours in the same amusements we are accustomed to enjoy when in London." The scientific spirit was willing, but the eighteenth-century flesh was weak. Whenever a few hours of sunshine "rendered it practicable to walk," however, they sallied forth with their microscopes, observed caterpillars, made observations upon snails, set down carefully for their readers (and perhaps for the Royal Society) observations, which if they found no immortal niche in the *Philosophical Transactions,* nevertheless provoked the minds of the female spectators to pondering upon the fullness and diversity of the universe, the power of a Deity Who had created it, and man's place as a middle link in the Scale of Being. Chance and the weather led the female spectators to the telescope as well as to the microscope, and their "night thoughts" were of the vastness of the visible universe. Both in its satire and its seriousness, *The Female Spectator* reflects the enthusiasm of ladies for the "new science." If there is nonsense and satire here, there is also growing awareness of ideas the new science was bringing into human thought and imagination.[47]

V

The Female Spectator is late enough to offer a composite of popular ideas which for more than half a century had been developed by the microscope. But the greatest example of the effect of the new instrument on literary imagination is found in an earlier work—the best of the microscopical satires, both because its author was the greatest satirist of the age, and because *Gulliver's Travels* is a classic illustration of the extent to which the new instrument affected the *technique* of an artist. In this respect, *Gulliver's Travels* is a complement to *Paradise Lost,* of which, in many other ways, it is the antithesis. As Milton produced a new kind of cosmic poetry, a drama of interstellar space which could not have been composed before the telescope, so *Gulliver's Travels* could not

---

[47] The feminism of the ladies is amusingly shown in a section on microscopical anatomy, II. 196 ff., in which the Female Spectator raises the question whether dissection of the brain has proved any fundamental difference between the sexes, with subsequent reflections upon possibilities of women's education.

have been written before the period of the microscope nor by a man who had not felt both the fascination and the repulsion of a new Nature shown by a new instrument.

Swift's following of the satiric themes we have already seen is most obvious in his descriptions of the "enormous barbarians" in the *Voyage to Brobdingnag*. He himself gave a clue to the technique when he described the farmer's wife suckling her child:

I must confess no object ever disgusted me so much as the sight of her monstrous breast, which I cannot tell what to compare with, so as to give the curious reader an idea of its bulk, shape, and colour. It stood prominent six foot, and could not be less than sixteen in circumference. The nipple was about half the bigness of my head, and the hue both of that and the dug so varified with spots, pimples, and freckles, that nothing could appear more nauseous. This made me reflect upon the fair skins of our English ladies, who appear so beautiful to us, only because they are of our own size, and their defects not to be seen but through a magnifying glass, where we find by experiment that the smoothest and whitest of skins look rough and coarse, and ill-coloured.

I remember when I was at Lilliput, the complexion of those diminutive people appeared to me the fairest in the world; and talking upon this subject with a person of learning there, who was an intimate friend of mine, he said that my face appeared much fairer and smoother when he looked on me from the ground, than it did upon a nearer view, when I took him up in my hand, and brought him close, which he confessed was at first a very shocking sight. He said he could discover great holes in my skin; that the stumps of my beard were ten times stronger than the bristles of a boar, and my complexion made up of several colours altogether disagreeable.

The same trick of magnification of anatomical details under the magnifying-glass appears in the passage in which Gulliver described the beggars. The "woman with a cancer in her breast," the "fellow with a wen in his neck" are less giants in a fairy tale, than pathological studies under the microscope in an English laboratory.[48] In the same passage Gulliver described that "familiar animal, the Louse":

The most hateful sight of all was the lice crawling on their clothes. I could distinctly see the limbs of these vermin with my naked eye, much better than those of an European louse, through my microscope, and their snouts with which they rooted like swine. They were the first I had ever beheld, and I should have been curious enough to dissect one of them, if I had proper instruments.

The flies that pestered the Brobdingnagians were only flies to them, but to Gulliver they were "as big as a Dunstable lark." "Their loathsome excrement or spawn" was clearly visible to Gulliver, "though not to the natives of that country, *whose large optics were not so acute as mine in viewing smaller objects.*" The wasps that attacked Gulliver were "as large as partridges"; their stings, which interested Dr. Lemuel Gulliver, the scientist, were "an inch and a half long, and as sharp as needles." Those were the stings he took back to Gresham College, a true son of his age, "collecting" for the Royal Society in outlandish parts of the universe.

It is not alone in such magnifications that Swift shows himself following the fashion of satirical writers upon the microscope. More subtle is the scene in which "three great scholars" of Brobdingnag study in detail Gulliver, the animalcule, seeking to determine its species. These might be English *virtuosi,* examining under the microscope Leeuwenhoek's "little animals," watching their curious motions, trying to determine whether these tiny twisting *spirilla* possessed life, whether they were mechanisms or organisms, and if organisms, to what order of nature they were to be assigned:

His Majesty sent for three great scholars who were then in their weekly waiting, according to the custom in that country. These

48 Cf. also Gulliver's account of the skins of the Maids of Honour, Chapter V, which seemed fair to the eyes of their own race, but to Gulliver "so coarse and uneven, so variously coloured, when I saw them near, with a mole here and there as broad as a trencher, and hairs hanging from it thicker than pack-threads."

gentlemen, after they had a while examined my shape with much nicety, were of different opinions concerning me. They all agreed that I could not be produced according to the regular laws of nature, because I was not framed with a capacity of preserving my life, either by swiftness, or climbing of trees, or digging holes in the earth. They observed by my teeth, which they viewed with great exactness, that I was a carnivorous animal; yet most quadrupeds being an overmatch for me, and field mice, with some others, too nimble, they could not imagine how I should be able to support myself, unless I fed upon snails and other insects, which they offered, by many learned arguments, to evince that I could not possibly do. One of these virtuosi seemed to think that I might be an embryo, or abortive birth. But this opinion was rejected by the other two, who observed my limbs to be perfect and finished, and that I had lived several years, as it was manifest from my beard, the stumps whereof they plainly discovered through a magnifying-glass. They would not allow me to be a dwarf, because my littleness was beyond all degrees of comparison; for the Queen's favorite dwarf, the smallest ever known in that kingdom, was near thirty foot high.

The conclusion of the "three great scholars" was reached by methods common to such "Baconians" who, Swift suggests wryly, for all their worship of the new method, achieved results as futile as those reached by old syllogistic logic:

After much debate, they concluded unanimously that I was only *relplum scalcath,* which is interpreted literally, *lusus naturae;* a determination exactly agreeable to the modern philosophy of Europe, whose professors, disdaining the old evasion of *occult causes,* whereby the followers of Aristotle endeavored in vain to disguise their ignorance, have invented this wonderful solution of all difficulties, to the unspeakable advancement of human knowledge.

Even more significant of the effect upon Swift's imagination of the "new science" is the technique he employs throughout the first two voyages. Swift's "perspective" is as characteristic as Milton's. We may find the trick of these two voyages in either the telescope or the microscope, or perhaps we might better say that Swift had in mind a simple "pocket perspective" such as Lemuel Gulliver always carried,[49] whose principle was that of our opera-glasses, one end magnifying, the other affording far sight. At all events, in both voyages we find the heritage of the microscope. Gulliver studied the Lilliputians as the *virtuosi* studied ant-hills, bee-hives, "Eels in Vinegar," blades of grass swarming with life. He was impressed by the exquisiteness of parts of the animalcules, the perfection of tiny mites. The more closely he observed them, the more he became aware of their beauty, their symmetry, their perfection, their adaptation to their uses in the scheme of things. Like the scientists who studied the world of minutiae through their glasses, he pondered these matters and became increasingly aware of his own grossness and disproportion. Perfect proportion existed in the little world; his own seemed uncouth. The little creatures had a sharpness of perception far superior to that of man. Theirs was the "microscopic eye," which could see in their own world what Gulliver could not perceive. Beyond their civilization might be still other races of animalcules as small to them as they to us. Gulliver's mind turned back to this later when he found himself among the Brobdingnagians: "It might have pleased fortune to let the Lilliputians find some nation, where the people were as diminutive with respect to them, as they were to me."

At first, like the early microscopists, Gulliver was enthralled

[49] The "metaphor of the telescope," as a clue to Swift's trick of perspective has usually been attributed to Walter Scott in his *Works of Swift*, XI. 8. As W. A. Eddy pointed out in *Gulliver's Travels: A Critical Study*, 1923, p. 146, the analogy was used in the contemporary French translation of Des Fontaines, 1727, xix–xx: "Dans ces deuc *Voyages*, il semble en quelque sorte considérer les hommes avec un Télescope. D'abord il tourne le verre objectif du côté de l'oeil et les voit par conséquent très-petits: C'est *Le Voyage de Lilliput*. Il retourne ensuite son Télescope, et alors il voit les hommes très-grands: C'est *Le Voyage de Brobdingnag*." Swift himself suggested the trick in his "Essay on Modern Education," which first appeared in the *Irish Intelligencer*, No. IX, 1728 (see *Satires and Personal Writings*, edited W. A. Eddy, p. 80), when he said of the "Young Gentleman": "If you should look at him in his Boyhood thro' the magnifying End of a Perspective and in his Manhood through the other, it would be impossible to spy any Difference."

by the perfection of a tiny universe, and saw in the Lilliputian animalcules only exquisite proportion of parts. As he continued to study them, he became aware that these were creatures of like passions with ourselves, who had their loves and hates, their quarrels, wars and treacheries.

Gulliver's reflections parallel the reflections of science in the half-century following the invention of the microscope. Optimism was the prevailing note of the scientists as a whole. In the "little world" they saw the perfection of nature even in the "secret recesses" and "most minute parts" of the universe, and found further evidence of the wisdom and goodness of God who could create perfection so abundantly. Yet among the serious writers of the period—sometimes in the scientists, more often among laymen—another note is heard. Growing realism gave pessimism powerful argument for the persistence everywhere in the vastly extended universe of such characteristics as pride, lust, fear, and selfishness. If the discoveries of the microscope made man aware of the exquisite proportion of nature, it brought sharply before him his own disproportion. As with the telescope, he felt in some moods securely placed by a careful Deity in a safe middle point between the infinitely great and the infinitely small; but in other moods, he swung between two extremes,

> Placed on this isthmus of a middle state,
> A being darkly wise and rudely great.

To Swift and Pope as to Pascal, man was the great paradox, a tiny Gulliver in the Land of Giants, a gross and disproportionate Gulliver walking with uneasy steps among Lilliputians who yet could bind him to earth, tied down by the hair of his own head, his greatness powerless before their subtle strength.

As the Lilliputians to Gulliver, so Gulliver to the Brobdingnagians; as he to them, so the Giants to him. All depends upon who holds the glass, and which lens is used. The Brobdingnagians who found tiny Gulliver in the field of corn seemed to him in size and terror as the *virtuosi* may have seemed to insects in the fields. "Seven monsters like himself came towards him with reaping-hooks in their hands, each hook about the largeness of six scythes." Are these mere laborers in a corn-field in Brobdingnag—or man as he must seem to those insects upon a leaf Fontenelle's Philosopher casually plucked to show a lady? Here but a moment ago in Lilliput was man as he seems to the world of minute life, man the crown and summit of creation, as Gulliver remembered himself in the other coun-try, "whose inhabitants looked upon me as the greatest prodigy that ever appeared in the world." But in another moment here is man as he may seem—shall we say?—to the dwellers on another planet, to which our world is but a twinkling star, and we the mites that crawl upon it. "What a mortification . . . to appear as inconsiderable in this nation as one single Lilli-putian would be amongst us." Is this Lemuel Gulliver, or is it Pascal's "reed that thinks," placed like Pope's Man upon a middle isthmus, between an infinity of greatness whose mag-nitude seemed beyond human understanding and an infinity of littleness that neither his sight nor mind could grasp? "Un-doubtedly," reflected Gulliver, "philosophers are in the right when they tell us, that nothing is great or little otherwise than by comparison." It was a truth man had always known. Yet never did he come to comprehend that truth as in the days when first the telescope and then the microscope con-founded his vision, when instruments made him feel himself now lord of creation, now gross, uncouth, disproportionate, a lonely mite crawling in a universe too vast for his comprehen-sion—when instruments, in short, showed him man, as Lemuel Gulliver found him and as Swift's contemporary described him, "the glory, jest, and riddle of the world."

VI

In Swift we see deepening into seriousness ideas that among earlier popular writers had been largely matters for jest, for raillery, for light-hearted satire. Often the microscope did no more than provide a theme for literary amusement. But more profound implications were stirring among serious writers of the seventeenth and eighteenth centuries.

Some possible effects of the microscope I can only suggest, since they lead into technical fields in which I have no competence. Yet if I cannot give an answer, I may raise a question: What part was played by the microscope in stimulating developments in the history of mathematics, particularly in connection with the infinitesimal calculus? In early writings of Leibniz and Newton—letters, commonplace books and articles—I have frequently noticed the interest of both in the microscope. Was it because of that mutual interest that each of them developed his theory of the infinitesimal calculus, priority in the discovery of which has divided their followers into opposed camps ever since that time? Historians of art have suggested the influence of the microscope upon painting. In the period of the telescope, as I have suggested, there was —whether by coincidence or influence—a new interest in perspective. In the period of the microscope, we find exquisite detail in the depiction of the minute, an even greater appreciation than in the past for the symmetry and perfection of tiny works of nature. And yet as we recall the expertness and craftsmanship of observers of nature in earlier periods, which so far as we know had not even the magnifying-glass, it is dangerous to suggest that the microscope did more than intensify techniques already well established. Possible effects of the microscope upon literary and philosophical imagination may, however, be suggested in more detail.

In prefaces to microscopical books of the late seventeenth century, there emerges a chapter in the "Battle of the Books," which in England always tended to be rather scientific than merely literary. The telescope and microscope became powerful weapons in the hands of the "moderns." [50] Argument drawn from them proved more embarrassing to the "ancients" than any other they were forced to answer. Indeed, the argument

[50] From the publication of Henry Power's *Experimental Philosophy* in 1664 until at least the end of the century, there is hardly a writer on the telescope or microscope who does not devote a section of his Preface, Introduction, or Conclusion to the quarrel, using the new instruments as irrefutable argument for the superiority of the moderns.

was unanswerable. The "ancients" had had nothing to correspond to these modern instruments that had discovered new worlds and prophesied new universes. "Dioptrical Glasses," as Henry Power said in 1664,[51] "are but a Modern Invention. Antiquity gives us not the least hint of them." Therefore the knowledge of nature in antiquity was necessarily incomplete and fragmentary, compared with the knowledge of nature in "modern" times. Before the invention of the telescope, the ancients had erred in their celestial hypotheses. Before the invention of the microscope they had been equally mistaken "in their nearer Observations of the Minute Bodies and smallest sort of Creatures about us," and had failed to see "how curiously the minute things of the world are wrought." Not only had the "moderns" succeeded where the "ancients" had failed, but Power, an ardent Baconian, declared that now that man possesses such "instruments and helps," there is no limit to his discoveries in the future. "Who can tell how far Mechanical Industry may prevail: for the process of Art is infinite, and who can set a *non-ultra* to her endeavors?"

The same arguments for the superiority of *modern* over *ancient* recur in Hooke's *Micrographia,* one of many treatises of seventeenth-century science that reads like a new version of Bacon's *Advancement of Learning* and *Novum Organum.* Power had suggested the superiority of modern men over Aristotle who, were he alive to use the new instruments, "might write a new History of Animals." Hooke insisted on the superiority of "moderns" over Adam himself, suggesting that the new artifices of man might in time compensate for man's fall, since instruments would permit him to rectify "the operations of the Sense, the Memory, and Reason," and thereby to regain in some measure the perfection earlier man had lost. In 1667 Sprat's *History of the Royal Society,* following the outline of the *Advancement of Learning* and *Novum Organum,* surveyed advances made in knowledge since the invention of instruments, and prophesied the future. "The

[51] This and other quotations from Power are from the unpaginated Preface to *Experimental Philosophy.*

Microscope alone," he said, "is enough to silence all opposers. . . . By the means of that excellent instrument, we have a far greater Number of different kinds of Things reveal'd to us, than were contained in the visible Universe before, and even this is not yet brought to Perfection." [52]

For a time after the publication of Sprat's *History*, the English quarrel between "ancient" and "modern" threatened to degenerate into a controversy over the Royal Society. Fontenelle's *Entretiens sur la Pluralité des Mondes*, and his *Digressions sur les Anciens et les Modernes* made it again a general quarrel. Although Sir William Temple's reply in his *Miscellanea* was chiefly concerned with literature and art, even he introduced a passage on science in his attempt to prove that the "moderns" must bow to the "ancients." The battle began anew, and later English works found their clinching arguments for modern superiority in the new instruments.

By 1692 it was neither necessary nor expedient to treat the "ancients" with so much tact and courtesy as earlier writers had showed. William Molyneux went into the fray with vigor. In his *Preface* to the *Dioptrica Nova*,[53] he prophesied the triumph of the modern experimental method. Completely on the side of the "moderns," he had no word of praise for the "ancients," nor did he look back either to Aristotle or to Adam. All that he had to say of the "old philosophy" was disparaging. Credit for advance belonged solely to the moderns, and had come about "in this last Age," first by the "generous Undertakings of the Philosophical Societies of Europe," and secondly by the experimental method, the great modern contribution to knowledge. It was particularly through such inventions as the telescope and microscope, Molyneux believed, that the advancement of learning had come about. With the microscope "our Contemplation may be endless" and we shall

[52] *History of the Royal Society*, 1772, pp. 284-5. The first edition appeared in 1667.

[53] *Dioptrica Nova. A Treatise of Dioptricks in Two Parts. By William Molyneux of Dublin, Esq., F.R.S.*, London, 1692. My quotations are from the second edition, of 1709, p. 280.

find "that the Contrivance of the Almighty Creator is as visible in the meanest Insect or Plant, as in the greatest Leviathan or strongest Oak. To touch upon all the Wonders this Instrument shews us would be infinite."

Most ardent among the champions of "modern" science was William Wotton, the "Moderno" satirized in King's *Dialogues of the Dead*. While Wotton took his point of departure particularly from Temple, his *Reflections upon Ancient and Modern Learning* showed his knowledge of Sprat, Fontenelle, Grew, Power and Hooke. It was impossible for the ancients to proceed far in science, he said, because they lacked the *instruments* by means of which the greatest modern discoveries have been made. Even Harvey's hypothesis of the circulation of the blood, rejected by many of his contemporaries, rested for proof upon such microscopical observations as those of Leeuwenhoek. In addition to clarification of earlier theories, Wotton found, the microscope had added greatly to general knowledge. Rapid advances had been made in study of the anatomy of man, of animals, of vegetables. Relationship between these three kingdoms appeared more and more clear, as microscopical observation proved "that Nature follows like Methods in all sorts of Animals." To Wotton the most important contribution of such new instruments as the telescope and microscope lay in the coherence and intelligibility they had shown to exist in the universe, a coherence and intelligibility unsuspected by the "ancients."

From this time on, it is so customary to find the triumph of the "moderns" taken for granted by scientific writers that a defense of the other side is a matter of interest. John Hill in his *Essay in Natural History and Philosophy* [54] believed that, with one exception, discoveries of recent times did not com-

---

[54] *Essays in Natural History and Philosophy. Containing a Series of Discoveries by the Assistance of the Microscope. By John Hill, M.D.*, London, 1752. Hill's criticism of the new science was that it was not sufficiently concerned with utility. Like the critics of Sir Nicholas Gimcrack, he felt that the ardor for "collecting" was diverting men's minds from the real purposes of science.

pensate for what man had lost. Yet even Hill excepted from his condemnation the "microscope of which I am so fond." It alone has led to discoveries of great importance. Through it, though through little else, "modern" man was superior to "ancient."

But on the whole the battle had been fought and the "moderns" had won. In the eighteenth century, the controversy tends to drop out of prefaces to scientific books, as other arguments become more important. The characteristic attitude of later writers may be found in one of the most popular scientific books of the age, Henry Baker's *Microscope Made Easy*.[55] In his Introduction Baker took for granted the advantages of modern times over ancient. "Having only their naked Eyes to trust to, they were uncapable of any great discoveries of this sort." We however have at our disposal constant means of improving and aiding the senses, none greater than the microscope. "Microscopes furnish us as it were with a new Sense, unfold the amazing Operations of Nature, and present us with Wonders unthought of by former Ages." Like his predecessors, Baker reiterates the idea that our imaginations are expanded and our thoughts made more sublime as we realize anew the benignity of Deity expressed in the fullness and diversity of the universe. "These are noble Discoveries, whereon a new Philosophy has been raised, that enlarges the Capacity of the human Soul, and furnishes a more just and sublime Idea than Mankind had before, of the Grandeur and Magnificence of Nature, and the infinite Power, Wisdom, and Goodness of Nature's Almighty Parent."

As the controversy over *ancient* and *modern* dies, the controversy over "Nature" and "Nature's Almighty Parent," perennial at all times, emerges with new force. Many of the arguments are old but something in the proof is new. The microscope led to a dual conception of Nature, and opened

[55] Baker's *Microscope Made Easy* (1742) and *Employment for the Microscope* (1753) were frequently reprinted and probably the most popular and usable handbooks of the day. My quotations are from the Preface to the third edition of *The Microscope Made Easy*, p. xiii.

an interesting chapter in the long controversy over *nature* and *art*. The upholders of *art* had a strong point in their insistence that the new discovery of Nature had been brought about by the *artifices* of glasses, man's invention, so that Sir Formals might flatter Nicholas Gimcracks by saying: "Nor do I doubt but your Genius will make Art equal, if not exceed Nature," while Sir Nicholas looked forward, as did many writers on the microscope, to an indefinite future, since "there's no stop to Art." Yet the argument of the adherents of *Nature* was equally valid. Two passages, written more than a century apart, will suggest the persistence of an idea that became a commonplace. In 1678 John Wilkins wrote: [56]

I cannot here omit the Observations which have been made in these later times since we have had the use and improvement of the Microscope, concerning that great difference which by the help of that doth appear, betwixt *natural* and *artificial* things. Whatever is Natural doth by that appear, adorned with all imaginable Elegance and Beauty. There are such inimitable Gildings and Embroideries in the smallest Seeds of Plants, but especially in the parts of Animals, in the Head or Eye of a small Fly; Such accurate Order and Symmetry in the frame of the most minute Creatures, a Lowse or a Mite, as no man were able to conceive without seeing of them. Whereas the most curious Works of Art, the sharpest and finest Needle doth appear a blunt rough Bar of Iron, coming from the Furnace of the Forge. The most accurate engravings or embrossments seem such rude bungling deformed Works, as if they had been done with a Mattock or a Trowel. So vast a difference is there betwixt the Skill of Nature, and the rudeness and imperfection of Art.

In 1787 George Adams, in his *Essays on the Microscope,* reflecting the same attitudes, showed that the argument had proceeded further.[57]

[56] *Of the Principles and Duties of Natural Religion,* 1678, p. 80. The passage was also used (quoted from another edition) by John Ray in his *Wisdom of God,* 1735, p. 58.

[57] *Essays on the Microscope* by the younger George Adams were first published in 1787; my quotation is from the second edition, of 1798, pp. 711–2.

After having particularized so many of the works of Nature, let us now pay some attention to those of Art. But what an humiliating contract shall we meet with! If our design in viewing objects by the microscope be to discover beauty, harmony, and perfection, it will be necessary to limit our inquiries to the former, happily alone sufficiently abundant; if, on the contrary, we are desirous of discovering deformity and imperfection, we must confine ourselves to the latter. Even those works of art that appear to the unassisted eye as decisive proofs of consummate skill in the workman, and which excite our admiration for their apparent neatness and accuracy, when brought to this test, exhibit their real state; and, consequently, tend but to display the inferiority of the most finished performance of the ablest artist, when put in competition with the glorious productions of nature. The finest works of the loom and of the needle, if exhibited with the microscope, prove so rude and coarse, that were they to appear thus to the naked eye, so far from affording delight to our belles, would be rejected with disgust. But the more we inquire into the works of nature, the more fully are we satisfied of their divine origin: in a flower, for instance, we see how fibres too minute for the unassisted sight are composed of others still more minute, till the primordial threads or first principles are utterly indiscernible; whilst the whole substance presents a celestial radiance in its colouring; with a richness so superior to silver or gold, as if it were intended for the cloathing of an angel, and we have the highest authority for asserting, that the greatest monarch of the Past in all his glory, was not arrayed like one of these.

Suggesting to his readers that they examine through the microscope a group of objects, usually considered triumphs of artifice—laces, threads, coins, and other delicately made objects—he declared: "An inspection of a few of the above articles only will clearly demonstrate that as in the moral and political world, so in the works of art, perfection is unattainable by mortal man."

The point of view is common for decades. Nehemiah Grew concluded a section of the Dedication of his *Anatomy of Plants* with the words: "Nature speaketh these things: the only true Pallas, wherewith it is treasonable for the most couriously handed Arachne to compare." [58] John Ray compared works of Art "of extraordinary fineness and subtility," admired and "purchased at a great Rate" with those far more exquisite "minute Machines endued with Life and Motion" discovered by Leeuwenhoek's glasses.[59] The elder George Adams discussed the *nature-art* controversy, proving "the Defects of human Art, when compared to those of Nature." [60] Henry Baker devoted a chapter of *The Microscope Made Easy* [61] to consideration of "the Works of Art and Nature compared together and considered. Not only are the works of Nature superior to those of Art, but the Divine Artist shows the perfection of His Art even more truly in small things than in great." Examine as we will the works of man's art "with a good Microscope, and we shall immediately be convinced, that the utmost Power of Art is only a concealment of Deformity, an Imposition upon our want of Sight. . . . The nearer we examine, the plainer we distinguish, the more we can discover of the Works of Nature, even in the least and meanest of her Productions, the more sensible we must be made of the Wisdom, Power, and Greatness of their Author. . . . Let us apply the Microscope where we will, nothing is to be found but Beauty and Perfection."

But a different aesthetic, leading to another metaphysic, might be deduced from the microscope. Richard Bentley said in one of his Boyle Lectures: [62]

[58] *Anatomy of Plants,* 1682.
[59] *The Wisdom of God Manifested in the Works of Creation* (first edition 1692?), 1735, pp. 166 ff.; see also pp. 59 ff.
[60] *Micrographia Illustrata,* 1747, pp. 240 ff.
[61] *Microscope Made Easy,* ed. cit., chap. li, pp. 292–300.
[62] *A Confutation of Atheism from the Structure and Origin of Human Bodies,* a Boyle lecture delivered in 1692, reprinted in *Sermons Preached at Boyle's Lectures . . . By Richard Bentley,* edited Alexander Dyce, London, 1738, pp. 58–9. In the next passage Boyle, continuing his argument for plan and design, makes a similar point about the sense of hearing: were that as acute as instruments may make it, man would be deafened and overwhelmed by the universe. I am inclined to think that this passage was

If the eye were so acute as to rival the finest microscopes and to discern the smallest hair upon the leg of a gnat, it would be a curse and not a blessing to us; it would make all things appear rugged and deformed: the most finely polished crystal would be uneven and rough; the sight of us our selves would affright us; the smoothest skin would be beset all over with ragged scales and bristly hairs; and besides, we could not see at one view above what is now the space of an inch, and it would take a considerable time to survey the then mountainous bulk of our own bodies. Such a faculty of sight, so disproportionate to our other senses and to the objects about us, would be very little better than blindness itself.

Bentley's is the mood of Algarotti's Lady when she observed through a microscope her skin that had been famed for its beauty. Gulliver was appalled by the ladies-in-waiting in Brobdingnag, as earlier the Lilliputians had shrunk back from the ugliness of a grossly exaggerated human being. Bentley's tone we shall hear again in a passage from the *Seasons* in which Thomson reflected the repulsion rather than delight one group of men found in the Nature disclosed by the microscope.

Equally interesting to the historian of ideas is another perennial controversy to which the microscope offered dual ammunition—the question whether Nature is characterized by *regularity* or by *variety*, whether the Great Artist expresses Himself in pattern, form, and proportion, or whether He pours himself forth with lavish and unrestraining hand. The apostles of *regularity* and *pattern* drew their arguments from the perfection of design in the markings of insects, the symmetry of their minute organs, the exquisiteness of parts. Works of art seen by Henry Baker under the microscope were "irregular," "rough," "jagged" and "uneven." But the silkworm's web delighted him by its perfection of design, the specks on the wings of a moth he found "accurately circular." George

----

Thomson's source for the passage (quoted below) from the *Seasons* which is out of key with the mood of Thomson's period, but more common in Bentley's.

Adams in the *Micrographia Illustrata* stressed "the proportion, regularity, and design of the infinitely small." One of the favorite proofs of the upholders of Nature in the aesthetic controversy came from consideration of "the Particles of Matter composing Salts and saline Substances," the extraordinary delicacy and intricacy of which fascinated the early microscopists. Henry Baker spoke for many when he said: [63]

That beautiful Order in which they arrange themselves and come together under the Eye, after being separated and set at Liberty by dissolution, is here described and shewn. Did they amongst themselves all compose but one kind of figure, however simple, with Constancy and Regularity, we should declare it wonderful: What must we then say, when we behold every Species working as it were on a different Plan, producing Cubes, Rhombs, Pyramids, Pentagons, Hexagons, Octagons, or some other curious Figures peculiar to itself; or composing a Variety of Ramifications, Lines and Angles, with a greater Mathematical Exactness than the most skilful Hand could draw them? . . . When therefore these Particles of Salts are seen to move in Rank and File, obedient to unalterable Laws, and compose regular and determined Figures, we must recur to that Almighty Wisdom and Power, which planned all the System of Nature, directs the Courses of the Heavens, and governs the whole Universe.

Yet the layman was impressed too by the *variety* of microscopical Nature. Fontenelle's Lady, studying the green leaf, was amazed to see "how Nature has formed variety in the several Worlds" and to learn that there are no two things exactly alike in the universe. Her fancy was "confounded with the infinite Number of living Creatures," her thoughts "strangely embarrass'd with the variety that one must of Necessity imagine to be amongst 'em; because I know Nature does not love Repetitions; and therefore they must all be different." The Female Spectators found, as "Philo-Naturae" had told them, that Nature was everywhere characterized by variety. Addison's dream-visitant had seen "millions of species"

[63] *Employment for the Microscope*, 1753, Preface, pp. iv–v. In his *Cosmologia Sacra* (1701) Nehemiah Grew came to the same point, pp. 14 ff.

subsisting on a green leaf, had "been shown a forest of numberless trees, which has been picked out of an acorn," and
had seen in vegetables, minerals, and metallic mixtures, the
"several kinds of animals that lie hid, and as it were lost in
such an endless fund of matter." Everywhere he had been
amazed at the profusion and variety of Nature. Nature was
becoming not the Nature of Milton's "Lady," whose

> full blessings would be well dispensed
> In unsuperfluous, even proportion,

but the Nature of "Comus," covering the earth, thronging
the seas, pouring

> her bounties forth
> With such a full and unwithdrawing hand.

The Nature that emerged from the microscope was a Nature
superabundant and lavish, prolific and creative. No leaf, no
stone, no blade of grass but swarms with inhabitants. Beyond
the eye, even beyond the microscope, man conceived a universe fuller, more diverse, perhaps even infinitely full. As
from the vastly extended universe displayed by the telescope
to the eyes of the earlier century, so from this vision of the
microscope, an occasional sensitive soul shrank back in fear
or loathing. Pascal turned from the vast spaces of the new
cosmic universe to a "prodigy equally astonishing," the world
of the minute. Morbidly his imagination dissected that world,
with its "limbs with joints, veins in these limbs, blood in these
veins, humors in this blood, globules in these humors, gases
in these globules." Beyond lay still "a new abyss . . . the conceivable immensity of nature, in the compass of this abbreviation of an atom." Here was a new "infinity of worlds, each
of which has its firmament, its planets, its earth . . . and on
this earth animals, and in fine mites, in which he will find
again what the first have given; and still finding in the others
the same thing, without end, and without repose, let him lose
himself in these wonders, as astonishing in their littleness as
the others in their magnitude. . . . Whoever shall thus con-

sider himself, will be frightened at himself." [64] This mood was momentarily reflected by James Thomson, who combined, in a long passage in *The Seasons* [65] the fascination and dread of the new Nature, with her "numerous kinds Evading even the microscopic eye."

> Full nature swarms with life; one wondrous mass
> Of animals, or atoms organized.

In the putrid streams of hoary fens, Thomson felt "the living cloud of Pestilence." The dank earth in the darkness of the forest "heaves" with animated life; the flower, the stone hold multitudes; the green-mantled pool conceals prolific life; the forests, the orchard, the "melting pulp of mellow fruit the nameless nations feed Of evanescent insects." Only Nature's concealment of her fearful superfluity saves man's reason:

> These, concealed
> By the kind art of forming Heaven, escape
> The grosser eye of man; for, if the worlds
> In worlds enclosed should on his senses burst,
> From cates ambrosial, and the nectared bowl,
> He would abhorrent turn, and in dead night,
> When silence sleeps o'er all, be stunned with noise.

Pascal's shrinking from a universe that had become too vast for human comprehension is not unusual in the seventeenth century. Thomson's momentary revulsion from a new universe of the microscopically small is almost unique in the eighteenth. More characteristic of the response of eighteenth-century poets was the *Night Thoughts* of Edward Young, who "accepted the universe" with exultation. Like his predecessor Henry More, Young said in effect:

> Then all the works of God with close embrace
> I dearly hug in my enlargéd arms.

Young's cosmic arms grew larger and larger. He grasped both the new heavens and a new earth, swarming with animalcules.

[64] *Thoughts . . . of Blaise Pascal,* translated O. W. Wight, 1893, pp. 158–60.
[65] "Summer," ll. 287–317.

He was pre-eminently the poet of the telescope, by which he sought to reach God's throne, yet he could also write of the microscope:

> Glasses, (that revelation to the sight!)
> Have they not let us deep in the disclose
> Of fine-spun Nature, exquisitely small,
> And though demonstrated, still ill-conceived? [66]

To Young, the discovery of the world of the minute was a logical corollary to earlier discovery of a new universe of the vast. Both were inevitable. Like his own Deity he summoned into being "with like ease a whole creation, and a single grain." Pope might say,

> Thro' world unnumbered though the God be known,
> 'Tis ours to trace him only in our own,

and draw back from the revelations of both telescope and microscope. Young opposed Pope here, as on many other issues. Young sought his God in the unnumbered worlds of cosmic universes and in the new infinity of world, of the microscope. His Deity expressed His overflowing goodness in a profusion of worlds and universes, great and small:

> Near, though remote! and though unfathom'd, felt!
> And though invisible for ever seen!
> And seen in all! the great and the minute;
> Each globe above, with its gigantic race,
> Each flower, each leaf, with its small people swarm'd,
> (Those puny Vouchers of Omnipotence)
> To the first thought, that asks, "From whence?" declare
> Their common Source. Thou Fountain running o'er
> In rivers of communicated joy! [67]

The telescopic universes of Galileo, the microscopic worlds of Leeuwenhoek and his followers were all Young's heritage. Though cosmic reaches remained the "wide pasture" of his imagination, he might find Deity everywhere:

[66] *Night Thoughts,* IX. 1575-8.
[67] *Ibid.,* IX. 2196-2204.

> True: all things speak a God; but in the small,
> Men trace out Him; in great he seizes Man;
> Seizes and elevates and wraps and fills
> With new inquiries, 'mid associates new.[68]

Optimism rather than pessimism was the dominant mood of one important chapter in the contribution of the microscope to thought: its aid in transforming the old Scale of Nature into the modern conception of evolution.[69] Sir Thomas Browne, discoursing upon spirits, expressed the old idea, as it had long been accepted.[70] "It is a riddle to me," he said, "how so many learned heads should so far forget their Metaphysicks and destroy the ladder and scale of creatures, as to question the existence of spirits." He continued:

For there is in this Universe a Stair, or manifest Scale of creatures, rising not disorderly, or in confusion, but with a comely method and proportion. Between creatures of meer existence, and things of life, there is a large disproportion of nature; between plants and animals or creatures of sense, a wider difference: between them and Man, a far greater; and if the proportion hold on, between Man and Angels there should be yet a greater.

Yet coexistent with the idea of a scale of nature had developed what Professor Lovejoy calls "the principle of plenitude," the conception of Nature as a *plenum formarum,* the assumption that the perfection of the divine essence "must constitute in a minutely graded hierarchy, a *continuum* of forms from highest to lowest, of which any two adjacent members differ only infinitesimally." To this assumption the microscope offered new proof. It proved that "there are no gaps in nature"; it began to show a Nature infinitely full and diverse. Unknown "stairs" or "steps" were discovered in "the ladder of being."

[68] *Ibid.,* IX. 772–5.

[69] Since my original study appeared, Professor Arthur Lovejoy's *Great Chain of Being* has made the history of the Scale of Nature and Chain of Being so familiar that I have omitted the section in which I discussed the background.

[70] *Religio Medici,* in *The Works of Sir Thomas Browne,* edited Geoffrey Keynes, 1928, I. 38–42.

The chain now included links long hidden from mortal eye. Lines of demarcation that had separated the "scale" of animals from the realm of vegetable life disappeared, as the microscope discovered more about the nature of each. Indeed as botanists and biologists continued their observation, a question rose whether the supposed line of demarcation between organic and inorganic matter was still a valid assumption. Fontenelle's Philosopher put into popular language a growing scientific belief, when he told the Marchioness that "several, even of the most solid Bodies, are nothing but an immense swarm of imperceptible Animals. . . . In short, every thing is animated, and the Stones upon Salisbury Plain are as much alive as a Hive of Bees."

Scientists and laymen were interested in every form of life in the scale of being, not only in the place of Man and God (a main theme of the theodicies of the period), but in the relationship of animals to man, to other animals, to plants. Gradually the scientists became convinced that *Natura non facit saltus.* There is no point at which animal life necessarily ceases and vegetable begins. Malpighi's studies of plant life, and Nehemiah Grew's discovery of the sex of plants marked a new chapter not only in biology but in philosophy and theology.

Grew published his first important observations in *The Anatomy of Vegetables* in 1672, but his most significant work appeared ten years later in *The Anatomy of Plants.*[71] In the Dedication of this volume Grew proposed his theory in short. Addressing the Patron of the Royal Society, he wrote:

Your Majesty will find, That there are *Terrae Incognitae* in Philosophy, as well as in Geography. And for so much, as lies here, it

[71] In the Preface to *The Anatomy of Plants,* 1682, Grew says that he began his work on vegetables and plants in 1664, and continued because of the encouragement of members of the Royal Society. His earlier work, *The Anatomy of Vegetables,* was presented to the Society on December 7, 1671. On the same day, he reported, they received a "manuscript (without Figures) from Seignior Malpighi, upon the same subject." Grew's first observations were made with the naked eye, his greatest discoveries only after his use of the microscope.

comes to pass, I know not how, even in this Inquisitive Age, that I am the first, who has given a Map of the Country.

Your Majesty will here see, That there are those things within a Plant, little less admirable, than within an Animal. That a Plant, as well as an Animal, is composed of several Organical Parts; some whereof may be called its Bowels. That every Plant hath Bowels of divers kinds, conteining divers kind of Liquors. That even a Plant lives partly upon Aer; for the reception whereof, it hath those Parts which are answerable to Lungs. So that a Plant is, as it were, an Animal in Quires; as an Animal is a Plant, or rather several Plants bound up into one Volume. . . . In sum, Your Majesty will find, that we come ashore into a new World, whereof we see no End.

The familiar passage in which Locke, in the *Essay on the Human Understanding*,[72] in 1690 suggested the significance of these ideas gave rise to much discussion and many implications. It is both possible and rational, Locke declared, that "there may be many species" both above and below us of which we have as yet no idea. In the visible or corporeal world, "we see no chasms or gaps." The descent "is by easy steps, and a continued series of things that in each remove differ very little one from the other." There are fishes, he pointed out, that have wings "and are not strangers to the airy regions," and birds whose blood is as cold as the blood of fishes, and flesh so like in taste that they may be eaten even by "the scrupulous" on holy days. There are animals so close to both birds and fish that "they are in the middle between both." The terrestrial and aquatic worlds are so closely linked together by amphibians—even by mermen, whose existence, Locke tells us, is "confidently reported"—that no one can say

[72] *An Essay concerning Human Understanding*, Book III, chap. vi, section 12, in *Works of John Locke*, edited J. A. St John, 1685, II. 49–50. Comparison of the later version of the *Essay* with the draft of 1671 shows that Locke had been following the microscopical investigations of his period. In the earlier version he was concerned—and that only in passing—merely with the *orders above* man. See *An Essay Concerning the Understanding, Knowledge and Opinion*, edited Benjamin Rand, Cambridge (Massachusetts), 1931, p. 162.

where one ceases and the other begins. The highest animals "possess as much knowledge and reason as some that are called men." And when we come to the supposed division between animal and vegetable, we shall find them so closely joined, that "there will scarce be perceived any great difference" between the highest of one and the lowest of the other; and so, concluded Locke, "till we come to the lowest and the most inorganical parts of matter, we shall find everywhere that the several species are linked together and differ but in almost insensible degrees." In 1698 Edward Tyson, the British "father of comparative anatomy," dissected a chimpanzee and discovered for the first time a "missing link" between man and animal much more convincing than Locke's hypothetical mermen. One lacuna was now filled; it was only a matter of time before scientists in their laboratories would discover others.

In Locke's passage, as critics have pointed out, lay some elements of the modern theory of evolution. For our purposes it is more important that we detect here the awareness of a Fellow of the Royal Society of possibilities suggested by the microscope. Locke was aware of similarities in function and structure shown by plants and animals. He believed that the microscope would prove the existence of other unknown and unsuspected species. There are no "gaps in nature." Life may continue "till we come to the lowest and most inorganical forms of matter."

Another philosopher emphasized the second aspect of this dual conception. Early writings of Leibniz—his letters and his contributions to periodicals—show that his imagination, too, had been stimulated by microscopical discoveries.[73] In

[73] References to Leeuwenhoek, Malpighi, Swammerdam and other microscopical observers are frequent in Leibniz's letters. An important early passage, showing nearly all his characteristic ideas, occurs in the "Système Nouveau de la Nature et de la Communication des Substances" published in the *Journal des Sçavans* in 1695. See particularly pp. 297 ff. for comments upon Leeuwenhoek and Malpighi. The best account of the effect of biological discovery on Leibniz seems to me John Dewey's in his introduction to the *New Essays Concerning Human Understanding*, pp. 33 ff.

Leibniz's comments upon Locke's passage, we find the same interest in the possibility that there are "no gaps in nature" and see even more clearly the seventeenth-century emphasis upon microscopic proof of the logical idea of nature as a *plenum formarum*. The predecessors of Leibniz were Plato and Plotinus, on the one hand, Grew and Malpighi on the other, but they were also such popular writers as Fontenelle, whose Philosopher pointed out to the Lady that "there be as many kinds of invisible as visible Creatures . . . an infinity of lesser Animals. . . . You will find the Earth swarms with inhabitants."

In the *Nouveaux Essais* Leibniz not only accepted Locke's contentions, but merged them with his own into a characteristic passage in which appear several of his dominant conceptions. He quoted Locke in a speech of "Philalethes," one of the characters in his dialogue, and replied, in the person of "Theophilus," raising the old question of certain "clever philosophers—*utrum detur vacuum formarum,* i.e., whether there are possible species, which, however, do not exist, and which nature may seem to have forgotten." This led him to his theory of "compossibles," that "all species are not compossible in the universe, great as it is." Granted this, Leibniz declared that "all things, which the perfect harmony of the universe can receive, exist therein." He discussed the possibility of "intermediate creatures between those which are far apart," and agreed that their existence was in conformity with the harmony of the universe he presupposed. His argument was based upon that *law of continuity* which "declares that nature leaves no gap in the order she follows." Here are several of Leibniz's most familiar ideas which he repeated frequently in succeeding years: his theory of "compossibles"; the *law of continuity* he had inherited from the past, which he was to make a principle of the universe; insistence upon the pre-established harmony, so peculiarly his that it has seemed original with him.

Locke's passage was quoted frequently in the years that followed, often with microscopical implications. Among the

English popularizers was Addison. While the "Spectator" had not hesitated to laugh at some tendencies of the microscopists, Addison was greatly impressed by the new ideas. In his essay of November 22, 1712, he suggested the interest anatomical studies of various forms of life had excited in thoughtful minds. He was impressed both by the repetition and the diversity of Nature. Let man consider animals, reptiles, insects, "and he will observe how many of the works of nature are published, if I may use the expression, in a variety of editions. . . . You will find the same creature that is drawn at large copied out in several proportions, and ending in miniature." That there is plan and design, not chance, in this repetition is clear, he declares, "if we apply it to every animal and insect within our knowledge, as well as to those numberless living creatures that are objects too minute for a human eye: and if we consider how the several species in the whole world of life resemble one another in very many particulars, so far as is convenient for their respective states of existence." In his essay of October 25 of the same year, Addison had commented upon Locke's passage. He recognized the obliteration of lines of demarcation between species, and also the vast extension of the *scale of nature*. He added details: "There are some living creatures," he says, "which are raised just above dead matter. . . . There are many other creatures but one remove from these, which have no other sense besides that of feeling and taste." This led him to consider "by what a gradual progress the world of life advances through a prodigious variety of species, before a creature is formed that is complete in all its senses." There are, too, different degrees of perfection in the senses enjoyed by animals, varying degrees of instinct "rising after the same manner, imperceptibly one above another, and receiving additional improvements according to the species in which they are implanted. . . . This progress in nature is so very gradual, that the most perfect of an inferior species comes very near to the most imperfect of that which is immediately above it."

Imaginatively Addison responded to the superabundance of

218

Nature in the universe of "little animals," reflecting in his language various of his predecessors:

It is amazing to consider the infinity of animals with which it [the material world] is stocked. Every part of matter is peopled; every green leaf swarms with inhabitants. There is scarce a single humour in the body of a man, or of any other animal, in which our glasses do not discover myriads of living creatures. The surface of animals is also covered with other animals, which are, in the same manner, the basis of other animals that live upon it; nay, we find in the most solid bodies, as in marble itself, innumerable cells and cavities that are crowded with such imperceptible inhabitants, as are too little for the naked eye to discover.

Some twenty years later, Pope in the *Essay on Man* combined conceptions gathered from Grew, Leeuwenhoek, Locke, Leibniz and Addison with older logical ideas into a system of the universe. Pope's "mighty maze" was certainly not without a plan. It is marked by coherence and Order:

> The gen'ral Order, since the whole began,
> Is kept by Nature, and is kept in Man.

We need not now consider his arguments about "Man," nor about those "orders above" him. But in the orders "below," we find the same awareness of *degrees* and *gradations* of sensation that the microscopical writers had stressed:

> Nature to these, without profusion, kind,
> The proper organs, proper pow'rs assign'd;
> Each seeming what compensated of course,
> Here with degrees of swiftness, there of force;
> All in exact proportion to the state;
> Nothing to add, and nothing to abate.
> Each beast, each insect, happy in its own.

Pope's succinct couplet echoes the microscopists who had had so much to say of difference in degrees of sensation and sharpness of sense between orders:

> Why has not Man a microscopic eye?
> For this plain reason, man is not a Fly.

His ethical conclusion repeats words familiar among the satirists,

> Say what the use, were finer optics giv'n
> T'inspect a mite, not comprehend the heav'n?

Pope's most familiar lines on the chain of being merely versify ideas which the scientific writers for half a century had been expressing in the prose prefaces to their works:

> See, thro' this air, this ocean, and this earth,
> All matter quick, and bursting into birth.
> Above, how high, progressive life may go!
> Around, how wide! how deep extend below!
> Vast chain of Being! which from God began,
> Natures ethereal, human, angel, man,
> Beast, bird, fish, insect, what no eye can see,
> No glass can reach; from infinite to thee,
> From thee to Nothing. . . .

In the new universe the scale of nature, once fixed and determined, was expanding beyond sight, though not beyond the possibility of proof. With Pope this may have been rhetoric; but to the scientific writers of the century it was *fact*, not to be disputed. Popular and religious writers took it equally for granted. The *Female Spectator* had much to say of man's attempts to "pry into the smallest works of the creation," as a result of which "new scenes of wonder every moment open to our eyes." John Ray, like other physico-theologists, devoted long sections of *The Wisdom of God* [74] to "the admirable Structure of the Bodies of Man and other Animals," sections in which he dealt with the world of the minute, the relationships between species, with the "multitude of species," and the perfection of "inferior creatures" both in their "Order" and "Degree."

Reflections of scientific writers upon this complex of ideas may be found in prefaces to many of the technical works of

[74] *The Wisdom of God,* ed. cit., pp. 367 ff.

science. John Turberville Needham devoted his preface less to "science" than to a consideration of what microscopical discoveries have done to the imagination of man: [75]

Nature under the Direction of its Creator, tho' prolifick beyond the reach of Imagination, and ever exerting its Fecundity in a successive Evolution of organised Bodies, boundless in Variety, as well as Number, has yet so much of Uniformity in all its Productions, that not only the Specifick Ascent, or Descent throughout the whole Scale of visible Beings is easy, and gentle by almost imperceptible Gradations; but also the Subordination of Worlds, if I may so term the several inhabited Portions of Matter, is concerned into a Harmony of Individuals as surprising, as that of the several Species, under which they are ranged.

In this Theory, which is more than a specious System, or a mere agreeable Sally of Imagination, since many Traces of it appear in Nature, a Drop of Water, the Diameter of which exceeds not a Line, may be a Sea, not only as daily Experience shows, in the Capacity which it has of containing, and affording Sustenance to Millions of Animals, but also in the Similitude which these very Animals may bear to several known Species in that part of the Creation, which is the Object of our naked Eyes.

If our Acquaintance with the Microscopical World could be extended beyond the Bounds which Nature has prescribed to it, or even was already carried as far as Observation may in process of Time advance it, the Truth of this Theory would, I believe, appear in a much stronger Light, than our present confined Experience can afford, tho' abundantly sufficient to clear it from the Imputation of a groundless Supposition: And yet imperfect as it is, it wants not Instances to prove, that the peculiar Inhabitants of several Portions of Matter often bear a near Resemblance to each other, tho' they differ extreamly in Magnitude. . . .

A microscopical Animal may therefore in Shape and relative Magnitude be to numberless Inferiors, what an Elephant, Ostrich, or Whale is in the several Kingdoms of Beasts, Birds, and Fish; and this in so extensive a Gradation, that the Descent in Scale of

[75] *An Account of Some New Microscopical Discoveries . . . Dedicated to the Royal Society* (by John Turberville Needham), London, 1745, pp. 1–4.

Beings is as boundless to our imagination, as its Ascent, on the one hand extending towards Immensity, on the other decreasing towards Nothing, ever approaching, for ever distant.

In 1742 we may find all these ideas brought together, with quotations from the passages of Locke and Addison, and reflections upon the whole cosmic scheme, in *The Microscope Made Easy* [76] of Henry Baker:

As the Microscope discovers almost every Drop of Water, every Blade of Grass, every Leaf, Flower, and Grain swarming with Inhabitants; all of which enjoy not only Life but Happiness; a thinking Mind can scarce forbear considering that Part of the Scale of Beings which descends, from himself, to the lowest of all sensitive Creatures, and may consequently be brought under his Examination.

It is wonderful to observe, by what a gradual Progression the World of Life advances through a prodigious Variety of Species, before a Creature is formed that is compleat in all its Senses: and, even amongst these, there is such a different Degree of Perfection in the Senses which one Animal enjoys beyond what appears in another, that tho' the Sense in different Animals be distinguish'd by the same common Denomination, it seems almost of a different Nature. If after this, we look into the several inward Perfections of Cunning and Sagacity, or what we generally call Instinct, we find them rising in the same Manner, imperceptibly, one above another, and receiving additional improvements according to the Species in which they are implanted.

This Progress in Nature is so very gradual, that the whole Chasm, from a Plant to a Man, is filled up with divers Kinds of Creatures, rising one over another, by such a gentle and easy Ascent, that the little Transitions and Deviations from one Species to another are almost insensible. And the intermediate Space is so well husbanded and managed, that there is scarce a Degree of Perception which does not appear in some one Part of the World of Life. Since then the Scale of Being advances by such regular Steps so high as Man, we may by Parity of Reason suppose, that it still proceeds gradually upwards thro' numberless Orders of Beings of a superior Nature to him: as there is an infinitely greater Space and Room for different

---

[76] *The Microscope Made Easy*, London, 1744, pp. 306 ff.

Degrees of Perfection between the Supreme Being and Man, than between Man and the most despicable Insect.

During the first century of the microscope there was emerging a new conception of both God and of Man. The new Deity is, in the first place, the Divine Artist, who draws in little as exquisitely as in large. His "Wisdom, Art, and Power . . . shines forth as visibly in the Structure of the Body of the minutest insect as in that of a Horse or an Elephant." [77] "The Contrivance of the Almighty Creator is as visible in the meanest Insect or Plant, as in the greatest Leviathan or strongest Oak." [78] Consideration of the intricacy of the minute leads one writer after another to rhetoric concerning the Creator of this marvelous universe. William Derham, in his Boyle lectures for 1711 and 1712, concluded each section of his analysis of the nature of man and of animals with a passage in praise of the Deity who had so miraculously created them. "To consider," he says after his long series of chapters on the structure of birds and insects,

that all these Things concur in minute Animals, even in the smallest Mite; yea, the Animalcules, that (without good Microscopes) escape our Sight; to consider, I say, that those minutest Animals have all the Joynts, Bones, Muscles, Tendons, and Nerves, necessary to that brisk and swift Motion that many of them have, is so stupendous a Piece of curious Art, as plainly manifesteth the Power and Wisdom of the infinite Contriver of those inimitable Fineries.[79]

Before His Art man stood abashed. As he looked at his own works through the new instrument, he realized as never before "the defects of human Art," but as he surveyed the most minute creations of God, "the more Excellencies and Mysteries appear; and the more we are enabled to discover the Weakness of our own Senses, as well as the Omnipotency and infinite

[77] John Ray, *Wisdom of God*, p. 180.

[78] William Molyneux, *Dioptrica Nova*, p. 280.

[79] William Derham, *Physico-Theology: or, A Demonstration of the Being and Attributes of God, from his Works of Creation* (fifth edition), 1720, pp. 366–7.

Perfections of the great Creator." [80] God alone may rejoice in an Art as perfect as Nature; for, in a sense more profound than even Sir Thomas Browne conceived, "Nature is the Art of God."

The God of the "new science" is, secondly, like created Nature, a Deity superabundant, prolific, pouring Himself forth with unrestraining hand, expressing everywhere His creative power in a world of minutiae, in a universe of innumerable species. "With what amazing Numbers has the Beneficence of the Creator, unlimited as his other Attributes, peopled and planted the Bottom of the Deep, where no human Eye looks into the Wonder of his Goodness, where no rational Creatures, no Heart capable of praising him for his Works, has an Acquaintance with them!" [81] To an earlier generation of telescopic observers, the heavens had declared the glory of God anew; to the microscopists, God exhibited Himself in the minute wonders of creation, more truly within their comprehension. They saw eternity in a grain of sand, a new infinity in a drop of water. They found God in even "mean and filthy things," as Bacon had foretold. The earliest English microscopical writer, pondering a mite through the microscope, found evidence of Deity:

> All things hee made of nothing, but in this
> hee made a thing that lesse then Nothing is. . . .
> This to our mind the a'theriall wisdome bringes
> how God is greatest in the least of things
> And in the smallest print we gather hence
> the World may Best read his omnipotence. [82]

Like Sir Thomas Browne who loved the "narrow engines" Robert Boyle found that "Wonder dwells not so much on Nature's Clocks as on her Watches." Henry Baker, pondering Boyle's words, went on to say: "Upon comparing the Structure

80 George Adams, *Micrographia Illustrata*, 1747, p. 242.

81 John Hill, *Essays on Natural History and Philosophy*, p. 21.

82 Henry Power, 'In Commendation of the Microscope," transcribed from a Sloane MS. in the British Museum by Thomas Bowles in *Isis*, XXI (1934), 73.

of a Mite with that of an Elephant, I believe we shall concur in the same Opinion. The Largeness and Strength of the One may strike us with Wonder and Terror, but we shall find ourselves quite lost in Amazement, if we attentively examine the several minute Parts of the Other." [83] The generation found God in what one called "the disregarded pieces and hustlement of the Creation." "You never enjoy the world aright," wrote Thomas Traherne, "till you see how a sand exhibiteth the wisdom and power of God." [84] Consideration of minute particles as the works of God offered a powerful argument against the "Epicureans" and "atheists" who hailed the revival of atomism. Those

> praty motes
> Far finer than the smallest groates
> Of sand or dust
> That swarm in sun,[85]

were tiny particles indeed to threaten orthodoxy. But to the microscopists there seemed reason to believe that ultimately even the atoms would be seen and studied, and that optic evidence would be brought to prove them, like the "Salts and Saline Substances," no "fortuitous concourse" but particles which, like the rest of the minute creation, obeyed the laws of Wisdom and Design.[86]

All this new world of the infinitely small, God had given man, and He had given him also the ability to discover it to

[83] Henry Baker, *Microscope Made Easy*, 1744, xv. He was referring to a passage in Boyle's *Usefulness of Experimental Philosophy*.

[84] *Centuries of Meditations*, Meditation 27.

[85] George Puttenham, *Partheniades* (Ballad Society), II. 82.

[86] Power wrote in the Preface to his *Experimental Philosophy* in 1664: "And, indeed, if the Dioptricks further prevail, and that darling Art could but perform what the Theorists in conical sections demonstrate, we might hope, ere long, to see the Magnetical Effluviums of the Loadstone, the Solary Atoms of light (or *globuli etererei* of the renowned Des-Cartes) the stringy particles of Air, the constant and tumultuary motion of the Atoms of all fluid Bodies, and those infinite, insensible Corpuscles. . . . And though these hopes be vastly hyperbolical, yet who can tell how far Mechanical Industry may prevail?"

His glory. The microscopists were certain that, far from God's objecting to their discovery of His works in these formerly hidden forms of nature, He had put into their hands the new instrument that they might know both His Power and Wisdom. "It is true," wrote George Adams,[87] "this instrument discovers to us as it were a new creation, new series of animals, new forests of vegetables; but he who gave being to these, gave us an understanding capable of inventing means to assist our organs in the discovery of their hidden beauties. He gave us eyes adapted to enlarge our ideas, and capable of comprehending a universe at one view, and consequently incapable of discerning those minute beings, with which he has peopled every atom of the universe. But then he gave properties and qualities to matter of a particular kind, by which it would procure us this advantage, and at the same time elevate the understanding from one degree of knowledge to another, till it was able to discover these assistances for our sight."

Why has Deity put within man's grasp this instrument, by means of which he possesses powers denied to man in the past? On the one hand, the microscopists answer, that man may praise God:

> Each moss,
> Each shell, each crawling insect, holds a rank
> Important in the plan of Him, who framed
> This scale of beings; holds a rank, which lost
> Would break the chain, and leave behind a gap
> Which Nature's self would rue. Almighty Being,
> Cause and support of all things, can I view
> These objects of my wonder; can I feel
> These fine sensations, and not think of Thee? [88]

Yet at the same time—and the growing temper of the age is reflected here—can man view these objects of his wonder, this chain of being, always existing, yet stretching from the in-

[87] *Essays on the Microscope*, 1798, p. 23.

[88] Benjamin Stillingfleet, "Poetical Effusion on the Oeconomy of Nature," in *Miscellaneous Tracts*, p. 126. It is quoted also by Adams as the climax of his passage given above.

finitely great to the infinitely small, and not praise Man?

The tributes to Deity of the microscopists are sincere and reverent; there is more than lip-service here. Yet behind them is an increasing awareness of vast new powers of Man. Fervent though the prefaces and conclusions of the microscopical volumes seem when they pay tribute to God, they become more fervent still when they pay tribute to Man. "In this system of Being," wrote Addison, "there is no creature so wonderful in its nature, and which so much deserves our particular attention, as man, who fills up the middle space between the animal and intellectual nature, the visible and invisible world, and is that link in the chain of beings which has been often termed the *nexus utriusque Mundi*." [89] Like their master Bacon, the microscopists genuflect, then seem to say: "And now having said my prayers, I turn to Men."

In the picture of Man as in the picture of Nature that emerged during the early period of the microscope, there was dualism. Optimism and pessimism were combined. "You have convinced my Reason," Fontenelle's Marchioness said to the Philosopher who showed her the new world of the microscope, "but you have confounded my Fancy." In that distinction is implied one explanation for the diverse responses of scientist and layman. Before the new universe discovered by telescope and microscope, "Fancy" sometimes stood aghast. As the new space led the poetic mind to ask again: "What is man that Thou art mindful of him?" so the teeming world of minute life. If microscopical dissection had proved that plants were much like animals and animals much like man, did they not also show the reverse? Was man but another animal, like in his destiny as in his structure? Was he too an automaton, a mere complex of parts, a "little world made cunningly," acting only by mechanical laws? Below him the long scale of nature stretched away indefinitely, perhaps infinitely; but what of his place in that scale? Was he in truth the crown of creation, or was he but a link in the chain, no more important than the fly with its "microscopic eye," the plant that proved

[89] *Spectator*, Number 519. The passage is also quoted by Adams.

"an animal in quires?" Was he, after all, the discoverer, even in one sense the creator of this new universe; or had he in his pride created a universe that might overwhelm its creator? He might discern the *contagium vivum* by means of glasses he had invented; could he thereby control those particles which, more dangerous than an army, silently pursued him? He had discovered a universe of life; would it prove a universe of death?

"The invention of Glasses," as Henry Baker pointed out, "brought under our Examination the two Extreams of Creation." [90] What of those extremes? "The Extreams of Great and Little as far as our Conception aided by Experience can trace them, are immensely distant from each other." [91] Between them man found himself incapable of comprehending either. "Our Ideas of Matter, Space, and Duration are meerly comparative, taken from Ourselves and Things around us, and limited to certain Bounds; beyond which, if we endeavor to extend them, they become very indistinct. The Beginnings and Endings, excessive Greatness or excessive Littleness of Things, are to us all Perplexity and Confusion." [92] This is the mood of Pope's Man, placed on the isthmus of a middle state, his greatness and his wisdom clouded by darkness; the mood too of Pascal's infinitely great and infinitely small, infinite series of universes repeating each other, until they stagger comprehension. His "reed that thinks" is yet a reed, at any moment to be overwhelmed by the universe. Thomson reflects this mood in his dank forests and fens, full of life, yet full of pestilence. Most of all this is the mood of Gulliver, terrified among the Brobdingnagians, his human dignity reduced to the level of a sport and plaything, a toy for a Brobdingnagian child, an insect to be as carelessly dispatched as one of the buzzing flies. This is the paradox of Gulliver,

90 *Microscope Made Easy*, p. xiv.

91 John Turberville Needham, *Account of Some New Microscopical Discoveries*, p. 3.

92 Henry Baker, *Microscope Made Easy*, pp. 300–301.

lonely among the Lilliputians, his greatness setting him apart as certainly from the world of the small as his smallness set him apart from the world of the great.

Even the scientist occasionally reflected these moods. There is a note of dread and warning in prefaces of scientific books that sounds in our ears today as if these prefaces had been written in our own times. Man has unleashed forces of nature; can he control them? Man may well be proud of his discoveries of human powers the "ancients" never possessed, but let him employ those powers in humility, for his invention may bring both life and death. Among these early scientists we hear a note that we ourselves have almost forgotten: the voice of jealous gods, and a Jealous God. "As I would not derogate from the Greatness and Eminency of Man," wrote Power in his early work,[93] "so I would not have him arrogate too much to himself. . . . Let us not therefore pride ourselves too much on the Lordship of the whole Universe." "In Pride, in reas'ning Pride, our error lies," wrote Pope. Seeing themselves as the crown of orders below them, men might forget the orders above. Yet their own discoveries, considered soberly, offered a correction to this danger. Comparison of the works of Nature and of Art should save them from the overweening pride by which the angels fell. "Such a Comparison," said Baker,[94] "must tend towards humbling the Self-conceit and Pride of Man by giving him a more reasonable and modest Opinion of himself." Not only that; consideration of the discoveries on which they prided themselves, should serve to show them the true mood of the discoverer: "Every Species of these Animalcules may also usefully serve to correct our Pride, and prove how inadequate our Notions are to the real Nature of Things; by making us sensible how little the larger or smaller Parts of the Creation could possibly be made for us; who are furnished with Organs capable of discerning to a certain Degree only of the great or little, all beyond which is as much

[93] *Experimental Philosophy*, pp. 162–4.
[94] *Microscope Made Easy*, p. 292.

unknown, as far beyond the Reach of our conception, as if it had never been." [95]

Yet we distort the prevailing mood of microscopical discovery if we read too much pessimism into the microscopical writers. There was warning, to be sure; but that warning arose from awareness of the greatness of man's discoveries, and an exultant feeling that he would not stop, but would—in reverence and humility—go on and on. The true scientific mood of this age is optimism. The "modern" glories in his accomplishments; human ingenuity has proved itself as never before. Of the qualities of man in which Derham finds his greatness in his "Survey of the Soul of Man," none is more important than "inventive Power." All that falls within the reach of his senses he has employed "to some Use and Purpose, for the World's Good." He has discovered the laws of the celestial bodies and made them serviceable to astronomy, navigation, geography. He has shown "noble Acumen," and "the vast Reach" of his soul in his invention of "nice and various instruments." "And lastly, What a wonderful Sagacity is shown in the Business of Optics, and particularly in the late Invention of the Telescope; wherewith new Wonders are discover'd among God's Works, in the Heavens, as there are here on Earth, with the Microscope, and other Glasses." [96]

Modern man has proved his superiority over all men of the past—even over our Father Adam. There enters into nearly all these works the "idea of progress." The future as well as the present is man's. His ingenuity will go on until he has discovered all that is hidden and remote—"and who can set a *non-ultra* to his endeavors?" From his consideration of the little world will emerge new wonder at the power of man and the greatness of his faculties. "If there be so divine vertue in parts that are so sordid and nothing considerable," says the author of the preface to Thomas Moufett's *Theatre of Insects*,[97] "how

[95] Baker, *Employment for the Microscope*, p. 230.

[96] *Physico-Theology*, 1720, pp. 260–5.

[97] Preface to *Theatre of Insects*, in Topsell, *History of Four-Footed Beasts*, 1658.

great may we suppose the Excellency of the same is which rules in heart and brain?" The scientific literature of the period echoes an exultation in the powers of man, a glory in the increase of his faculties and his comprehension. His imagination expanded with the expansion of the universe. "The Use of the Microscope," says Henry Baker,[98]

will raise our Reflections from a Mite to a Whale, from a Grain of Sand to the Globe whereon we live, thence to the Suns and Planets; and, perhaps, onwards still to the fixt Stars and the revolving Orbs they enlighten, where we shall be lost amongst Suns and Worlds in the Immensity and Magnificence of Nature.

The microscope leads "to a Discovery of a thousand Wonders in the Works of his hand who created ourselves"; it "improves the Faculties and exalts our Comprehension." [99] Upon its "noble Discoveries . . . a new Philosophy has been raised, that enlarges the Capacity of the human Soul." [100] Man exults in the fact that so much remains to be discovered. "There remains a boundless Scope for our Enquiries and Discoveries: and every Step we take serves to enlarge our Capacities." [101] With his eyes upon the vision of an indefinite future, Man strides forward, sure of his power, given by God as he is careful to acknowledge, but developed by Man.

It was natural that the microscope led more certainly than the telescope to optimism and fervent praise of Man. Before the telescopic vision of the cosmos, even a brave man might shrink back, appalled at immensity, lonely before infinity. But the material of the microscopists was at once intelligible and flattering to man's sense of superiority. His attention turned more and more from cosmic space and from the "orders" above that emphasized his littleness to the worlds beneath him, to whom he was indeed a monarch, a Colossus. The swarming inhabitants of Fontenelle's green leaf, the expanded scale of

[98] *Microscope Made Easy*, p. 300.
[99] John Hill, *Essays in Natural History and Philosophy*, 1752, p. 6.
[100] Henry Baker, *Microscope Made Easy*, p. xiii.
[101] *Ibid.*, p. 310.

nature that included bird-fishes, Tyson's ape, and even Locke's mermaid—all these were man's inferiors: compared with them he *was* the climax of creation. In his ability to discover and to know these lower orders, to realize their dependence upon him and his power over them, lay his greatness. "This is the highest Pitch of human Reason," wrote Thomas Sprat, in Horatian mood with microscopic proof, "to follow all the Links of this Chain, till all their Secrets are open to our Minds; and their Works advanc'd, or imitated by our Hands. . . . This is truly to command the World; to rank all the Varieties and Degrees of Things, so orderly one upon another, that standing on the Top of them, we may perfectly behold all that are below, and make them all serviceable to the Quiet, and Peace, and Plenty of Man's Life." [102]

Sprat's words in 1667 are an anticipation. Henry Baker's, three-quarters of a century later, mark a conclusion of a period of discovery of Nature, of God, and of Man:

Every Creature is confined to a certain Measure of Space, and its Observation stinted to a certain Number of Objects: but some move and act in a Sphere of wider Circumference than that of others, according as they rise above one another in the Scale of Existence. This Earth is the Spot appointed for Man to dwell and act upon: he stands foremost of all the Creatures here, and links together Intelligence and Brutes. The Sphere of his bodily Action is limited, confined and narrow; but that of his Mind is vast, and extensive beyond the Bounds of Matter. Form'd for the Enjoyment of intellectual Pleasures, his Happiness arises from his Knowledge; and his Knowledge increases in Proportion as he discovers and contemplates the Variety, Order, Beauty, and Perfection of the Works of Nature: whatever, therefore can assist him in extending his Observations, is to be valued, as in the same Degree conducive to his Happiness.

What we know at present, even of Things the most near and familiar to us, is so little in comparison of what we know not, that there remains a boundless Scope for our Enquires and Discoveries; and every Step we take, serves to enlarge our Capacities, and give us

[102] *History of the Royal Society*, 1722, p. 384.

still more noble and just Ideas of the Power, Wisdom, and Goodness of the Deity.[103]

In the conclusion of this passage, we have a momentary vision of the heavenly city of eighteenth-century microscopists, who had discovered a universe full of infinite variety, who had experienced the fascination of Leeuwenhoek as he watched his millions of "little animals," who felt themselves surrounded by a cloud of witnesses, all partakers with them of the overflowing creative power of a Deity which, like themselves, had no desire to restrain that power:

The Universe is so full of Wonders, that perhaps Eternity alone can be sufficient to survey and admire them all: perhaps, too, this delightful Employment may be one great Part of the Felicity of the Blessed. When the Soul shall become divested of Flesh, the Pleasures of Senses can be no more: and if, by a continued Habit, any Longings after them shall hang about it, such Longings must create a proportionable Degree of Wretchedness, as they can never possibly be gratified. But if its principal delight has been in the Contemplation of the Beauties of the Creation, and the Adoration of their Almighty Creator, it soars, when disembodied, into the celestial Regions, duely prepared for the full Enjoyment of intellectual Happiness.

The mood of the microscopists is not that of classical restraint, of moderation, of acceptance of human limitation. These are insatiable romanticists, to whom limitation is not an answer but a challenge, who find one life too short for man's accomplishment, and feel man's greatness in his refusal to be content with less than all. Like that "little She-Philosopher," who preferred her microscope to her lover, they can conceive no life in the future more perfect than one in which they will continue to discover the wonders of Nature.[104] Their Utopia

103 *Microscope Made Easy*, pp. 309-310.
104 Important implications which escaped me when I was writing this monograph have since been developed by Ernest Lee Tuveson, *Millennium and Utopia*, Berkeley, 1949.

is no heaven city of perpetual calm, but an Eternity in which they may expect through infinite time the delights of discovery, of accomplishment, of fuller and fuller (though never complete) contemplation of the beauties of Nature, and—one likes to believe—celestial "Elaboratories" with perfect microscopes.

Index

235